ISBN 978-1-330-25939-9
PIBN 10004400

1 MONTH OF
FREE
READING

at

www.ForgottenBooks.com

By purchasing this book you are eligible for one month membership to ForgottenBooks.com, giving you unlimited access to our entire collection of over 1,000,000 titles via our web site and mobile apps.

To claim your free month visit:

www.forgottenbooks.com/free4400

English
Français
Deutsche
Italiano
Español
Português

www.forgottenbooks.com

Mythology Photography **Fiction**
Fishing Christianity **Art** Cooking
Essays Buddhism Freemasonry
Medicine **Biology** Music **Ancient**
Egypt Evolution Carpentry Physics
Dance Geology **Mathematics** Fitness
Shakespeare **Folklore** Yoga Marketing
Confidence Immortality Biographies
Poetry **Psychology** Witchcraft
Electronics Chemistry History **Law**
Accounting **Philosophy** Anthropology
Alchemy Drama Quantum Mechanics
Atheism Sexual Health **Ancient History**
Entrepreneurship Languages Sport
Paleontology Needlework Islam
Metaphysics Investment Archaeology
Parenting Statistics Criminology
Motivational

LESSONS

IN

CHÉMISTRY.

BY

WILLIAM H. GREENE, M.D.,

PROFESSOR OF CHEMISTRY IN THE PHILADELPHIA CENTRAL HIGH SCHOOL, ETC.

PHILADELPHIA:

J. B. LIPPINCOTT & CO.

LONDON: 15 RUSSELL STREET, COVENT GARDEN.

1884.

Copyright, 1884, by J. B. LIPPINCOTT & Co.

TABLE OF CONTENTS.

237556

ADVICE TO TEACHERS.

THE object of a limited course in chemistry is not to make chemists of the pupils, but to teach them what chemistry is, what it has accomplished, and what it may accomplish. The study of science can be made attractive only by arousing natural curiosity as to the cause of natural phenomena, and no greater mistake can be committed than to endeavor to make the facts of chemistry dependent upon its theory. It is true that here and there an exceptional pupil may grasp that theory and acquire the science; but even in such cases no special interest is developed, while to the more practical the subject becomes both foolishness and a stumbling-block.

The successful teacher of chemistry is not only thoroughly familiar with his science; he loves it. It is not enough that he has read several text-books on chemistry; he must be practically acquainted with all the phases of the facts with which he deals, and must have at least a general knowledge of the literature of the subject. His endeavor will then be to impart to his pupils some part of his own enthusiasm.

Chemistry is peculiarly a study of observation, and it should be taught as it has been developed,—first by the careful examination of facts, then by the theoretical explanations suggested by those facts. By new experiments the interest of the pupil is at once awakened, and will not flag during the consideration of the theory which explains the experiments.

The pupil is not to suppose that certain compounds are without importance, but is to understand rather that only the more important and more practical can be considered in the time devoted to study.

While the time has passed when valuable research can be conducted with inexpensive apparatus, a comparatively extended course on chemistry can be illustrated at little cost. Every teacher of chemistry should have some knowledge of glass-blowing, and some mechanical ingenuity to adapt the same apparatus to various purposes. Such skill is readily acquired by practice.

3225 SANSOM STREET, PHILADELPHIA,
1st May, 1884.

1*

5

THE DECIMAL SYSTEM OF WEIGHTS AND MEASURES, AND THE CENTIGRADE SCALE OF THE THERMOMETER, ARE USED IN THIS BOOK.

1 Metre   = 39.370708 inches.
1 Centimetre = 0.39370   "
1 Millimetre = 0.03937   "

1 Inch   = 2.539954 centimetres.

	OUNCES TROY = 480 GRAINS.	POUNDS AVOIRDUPOIS.
1 Milligramme =	0.000032	0.0000022
1 Centigramme =	0.000321	0.0000220
1 Decigramme =	0 003215	0.0002204
1 Gramme =	0 032150	0 0022046
1 Decagramme =	0.321507	0.0220462
1 Hectogramme =	3.215072	0.2204621
1 Kilogramme =	32.150726	2.2046212

1 Gramme = 15.4323 grains.

1 Grain   = 0.064799 grammes.
1 Oz Troy   = 31 103496   "
1 Lb. Avoirdupois = 0.453495 kilogrammes.

1 Cubic Centimetre of water at 0° weighs 1 gramme.

To convert centigrade degrees into Fahrenheit degrees, multiply by 9, divide by 5, and add 32°.

To convert Fahrenheit degrees into centigrade degrees, subtract 32°, then multiply by 5, and divide by 9.

CENTIGRADE SCALE.	FAHRENHEIT SCALE.	
300°	572°	
200°	360°	
150°	302°	
100°	212°	Water boils.
90°	194°	
80°	176°	
70°	158°	
60°	140°	
50°	122°	
40°	104°	
30°	86°	
20°	68°	
10°	50°	
0°	32°	Water freezes.
--10°	14°	
—17.8°	0°	
—20°	—2°	
—30°	—22°	
—40°	—40°	Mercury freezes.

A dull red heat is about 500° centigrade, or 950° Fahrenheit.

A high red heat is about 1000° C., and a white heat about 1500°.

LESSONS IN CHEMISTRY.

LESSON I.

INTRODUCTION.

Chemistry is the science which studies the differences of different kinds of matter.

1. Substance.—Matter occupies space, and can be measured and weighed. The different kinds of matter constitute so many different substances, which are distinguished from one another by general properties, such as color, relative weight, hardness, etc., and no two substances can be alike in all properties.

Some substances are capable of existing in the three possible states, as solid, liquid, and gas. Water is the most common example of such a substance; by the action of more or less heat it can be converted at pleasure into steam, liquid water, or ice. However, if we strongly heat a piece of wood or some sugar, these substances will not be melted into liquids or changed to vapor, but will be transformed into entirely different kinds of matter, from which we cannot again obtain the original substance.

2. Physical Changes.—We all know that water is capable of existing in many forms: mist, fog, rain, frost, snow, sleet, and ice all represent the same substance, and we know that these forms may change one into another while the nature of the substance remains unaffected. The salt water of the ocean differs from the fresh water of the rivers which flow into it, only because the sea-water contains salt and other forms of matter; but these substances are not water. If salt water be boiled, and a plate or other cold surface be held in the steam given off, drops of water will condense on the plate, and will be found to be perfectly fresh. The changes which convert ice into water, or water into steam or ice, are called physical changes, because the nature of the substance is not affected; a little more heat, or a little less, changes the water into steam or ice, or the steam or ice back again to water.

3. **Chemical Changes.**—Water may, however, undergo other changes, in which its nature is altered and new substances are produced.

Let us fill a test-tube with water, and, closing the open end with the thumb, invert it in a small vessel of water. Then we wrap a morsel of sodium (see § 414) about as large as a pea, in a small piece of wire gauze, twist around this a wire which may serve as a handle, and now, raising the tube so that its mouth is just below the surface of the water, we push the gauze under the edge of the tube (Fig. 1). A small piece of sodium must be used,

FIG. 1.

for the experiment is often ended by a little explosion, which might break the tube if a large piece were taken. As soon as the water touches the sodium, bubbles of gas rise to the surface and are collected in the tube. If the latter be not quite filled, the gauze may be withdrawn, *perfectly dried* by holding it in a flame, and another piece of sodium introduced, so that the tube shall be quite filled with gas. When this is accomplished, we may raise the tube from the water, still carefully holding it bottom upwards, and on introducing into it a lighted match or taper the gas will take fire, but will extinguish the taper, which will, however, be relighted by the burning gas as it is withdrawn (Fig. 2).

We shall presently learn that this gas, which is called hydrogen, does not come from the sodium, nor does it contain any sodium; it must therefore be produced from the water, and that portion of the latter which has yielded the hydrogen must be completely altered in its nature. The change is called a chemical reaction.

4. **Elements and Compounds.**—If water is thus capable of yielding another substance produced wholly from the water, we must believe that water is a compound substance, composed of more than one kind of matter. Of the innumerable forms of matter with which we are familiar, all, excepting comparatively few, are compounds, and by various means may be converted into simpler forms.

FIG. 2.

Chemists are acquainted with about sixty-eight substances which they have been unable to change into more simple forms. These substances are called elements.

5. Mercuric oxide is a heavy, red powder. We introduce a small quantity of this powder into a test-tube, and heat it in the flame of a spirit-lamp or Bunsen burner (Fig. 3). We will first notice that its color darkens; but this change is only physical, for if we remove the tube and allow it to cool, the original color is restored. If, however, we continue to heat it, in a short time a bright mirror forms on the glass in the cooler part of the tube: we now light a match, allow it to burn for a moment, and then blow it out, so that it may still retain a spark of fire; on introducing this spark into the tube it at once bursts into flame, and the match is relighted. The experiment may be repeated a num-

ber of times, extinguishing the match and relighting it. When we have sufficiently studied this phenomenon, we may examine the tube, and we will find that the mirror in the interior is composed of little globules of metallic mercury; we may shake them out and unite them in one. The gas which has been given off, and which causes such brilliant combustion, is called oxygen. The mercuric oxide was a compound body. By the aid of heat we have separated it or decomposed it into two other substances, —mercury and oxygen. Mercury and oxygen are elements; chemists have not been able to convert either of them into other substances of simpler nature.

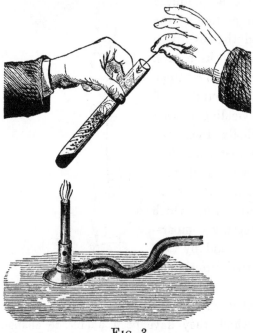

FIG. 3.

6. Sulphur and copper are also elements. We all know the yellow color and brittleness of sulphur, and the red color and flexibility of copper. We will put into a test-tube like that used in the last experiment a few small pieces of sulphur, and on top of them some copper turnings or a bunch of copper wire. We heat the tube; the sulphur melts, and presently begins to boil; but in a few moments we notice that the copper becomes very hot, much hotter than the portion of the tube which contains only sulphur. A chemical phenomenon is taking place, and the chemical action develops great heat. When the experiment has terminated, we allow the tube to cool, break it, and find that it no longer contains copper, and unless we have used too much sulphur we will find that the latter also has disappeared. In the place of the sulphur and copper there is a black,

brittle substance which results from the chemical union of the two elements. This substance is called copper sulphide.

In our first experiment we have seen the *decomposition* of a compound ; in the second we have caused the *combination* of two elements.

If we were to carefully collect and weigh all of the oxygen and mercury, we would find the weight exactly equal to that of the mercuric oxide from which they were obtained. If we make the second experiment in a long tube so that no sulphur vapor may escape, the weight of the tube will be the same whether it contain sulphur and copper or the copper sulphide resulting from their union. Nothing is lost or gained in either combination or decomposition.

7. Chemical Combination is not Mixture.—Mercuric oxide is not a mixture of oxygen and mercury, nor is copper sulphide a mixture of copper and sulphur. We may grind the last two substances to the finest powders and mix them together, but in this mixture we can by the aid of a microscope distinguish the particles of each substance. No microscope would enable us to detect sulphur and copper in the black copper sulphide. Since, however, we can separate the oxygen from the mercury, as we have seen, and the sulphur from the copper, we must believe that the elements still exist in their compounds. What is, then, their real condition ?

8. Molecules.—We know that under certain conditions any substance may change its volume. When a piece of iron is heated it grows larger ; when it is cooled it becomes smaller. We cannot believe that the matter of the iron actually increases in size when it is warmed, although it occupies a greater volume. We can understand this change in volume by believing that there are in the iron spaces or pores which increase in size when the metal expands, and grow smaller when it contracts. These pores must be very small, for we cannot perceive them by the aid of the most powerful microscopes. All substances must be porous, and we can satisfy ourselves by a simple experiment.

We have a glass tube about a foot long, closed at one end, and near the other blown out in two bulbs, and the part between the bulbs is rather narrow (Fig. 4). We pour water into this tube until it is filled to the top of the lower bulb. The water has been recently boiled, to drive out all of the air which was dissolved in it. We then fill the remainder of the tube with alcohol, and cork it tightly; the alcohol, which we have colored with a little aniline dye, does not at once mix with the water, because the latter is the heavier. We now invert the tube, and we see the lighter alcohol rise through the water, and at the same time the two liquids become thoroughly mixed. But as they mix we see little bubbles forming, and there is presently a small empty space at the top of the tube. This space is not filled with air; for if we put the mouth of the tube under water and draw the cork, the water will rise and fill the tube. The mixture does not then occupy as much volume as the substances before mixture. We must explain the experiment by saying that the water and alcohol are porous, and that the water runs into the pores of the alcohol, and the alcohol into the pores of the water.

If substances be in this manner porous, they must consist of small particles which are separated from one another by spaces. Both spaces and particles are so small that we can never hope to see them, but we have reason to believe that the spaces are quite large in comparison to the particles of matter which they separate. These particles are called *molecules*, and for our purposes of study we may consider that the spaces between them are perfectly empty.

Fig. 4.

9. **Atoms and Molecules.**—The little particles of which a chemical substance is composed are called molecules, and we shall learn reasons for believing that all molecules of the same kind, that is, of the same substance, have the same size and weight.

The kind of matter in a molecule must be the same as that in any quantity of the substance. If from mercury we can obtain only mercury, the molecules of mercury must consist of that element only; but if from mercuric oxide we can obtain both mercury and oxygen, the molecules of mercuric oxide must contain both of those elements. In that case there must be particles still smaller than molecules, and to these smallest particles we give the name *atoms*. Since chemists cannot separate oxygen into any other substances, we believe that the atoms of oxygen are unalterable by any known force, be it physical or chemical; and it is the same with the atoms of all other elements.

10. The atom is then the ultimate particle of matter, and the nature of an element will depend upon the nature of its atoms.

11. A molecule may consist of one or of several atoms: in the latter case, if the atoms be of the same kind the molecule will be that of an element or simple body, but if they be of different kinds the molecule must be that of a compound. The molecules of hydrogen can contain only atoms of hydrogen, and the molecules of oxygen must consist of atoms of oxygen only, but the molecules of water contain atoms of both hydrogen and oxygen; those of copper sulphide contain atoms of sulphur and of copper.

The nature of a substance will then depend upon the number and kind of atoms contained in its molecules. We have seen that mercuric oxide contains mercury and oxygen: let us pour a little nitric acid on some of this red powder contained in a test-tube, and warm the mixture over a lamp. The mercuric oxide disappears, and we obtain a colorless liquid; if we pour this liquid into a flat dish, and set it aside in a warm place, we will find after a time a mass of white crystals. A chemical change has taken place: while dissolving in the nitric acid, the mercuric oxide has been converted into mercuric nitrate. The latter body contains mercury, oxygen, and nitrogen, and its molecule must consist of atoms of each of those elements: the new molecule is more complex than that of mercuric oxide, which contained only two kinds of atoms.

12. **Chemical Affinity.**—The force with which atoms are held together is called *affinity*. Its energy is not the same for all atoms, and depends on many conditions. While heat may aid in the formation of a compound, as with copper sulphide, it may determine decomposition, as with mercuric oxide. Other forces,

2

light and electricity, may act in the same manner, in one case producing combination, and in another decomposition: in every case
the result depends upon the energy with which the atoms of the
molecule are held together. Why must we heat the copper and
sulphur before they will combine? Simply because the atoms of
sulphur hold strongly together in the molecules of sulphur, and
the atoms of copper in the molecules of that metal: we must
therefore communicate to these molecules so much energy in the
form of heat that the atoms of sulphur may be sufficiently loosed
from each other to catch hold of the atoms of copper. Then instead of molecules containing atoms of sulphur or copper only, we
have others containing sulphur and copper.

13. We mix a small quantity of powdered cupric oxide, a black
compound containing only copper and oxygen, with about one-
seventh its weight of powdered charcoal, and heat the mixture
in a test-tube. When the mixture becomes hot, we see that
the black color changes to reddish brown: after the powder has
cooled, we turn it out, and find that it is very finely divided
copper. We cannot heat the cupric oxide alone hot enough to
decompose it; but the charcoal, which is an element, has a strong
affinity for the oxygen, and easily takes it away from the copper.

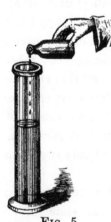

The charcoal and oxygen combine together and
form a gas called carbon dioxide, which passes
out of the tube. We can prove that some
new gas is formed during the experiment, for
if we push a lighted match into the mouth of
the tube while it is being heated, the flame
will be extinguished, an effect exactly opposite to that which was produced while heating the mercuric oxide. We then explain the
experiment, which is at the same time a combination and a decomposition, by saying that
the oxygen has a stronger affinity for the charcoal than for the copper.

FIG. 5.

14. Into a jar or glass nearly filled with water (Fig. 5) we
pour first a few drops of a solution of mercuric chloride, and then

some solution of potassium iodide. At once a pink or red *precipitate* is formed, showing that a chemical change has taken place. Both mercuric chloride and potassium iodide are color-less: the first contains two elements, mercury and chlorine, while the second also contains two, potassium and iodine. The chemical change takes place because the affinity of potassium for chlorine, and of mercury for iodine, is stronger than that of potassium for iodine and of mercury for chlorine. Consequently both of the original substances are decomposed, and *two* new substances are formed, mercuric iodide, which is insoluble (unless too much of either substance has been used), and potassium chloride, which remains dissolved in the liquid. This is an example of *double decomposition*, the most common kind of chemical change. A comparison will show that it closely resembles the double decomposition between sulphur molecules and copper molecules already explained (§ 12).

15. Chemical affinity is not to be regarded as a special force, but only as one form of energy; it is manifested between atoms, which it holds together in the molecules, just as these molecules are held together by the force of cohesion. Affinity depends not only on the kind of atoms between which it is exerted, but on the temperature: elements which have strong affinities for each other at a given temperature may not manifest any such affinity at other temperatures.

16. **Metals and Non-Metallic Elements.**—For convenience of study the elements are divided into two general classes, metals and non metals. The reasons for which an element is considered to be a metal or a non-metal can be understood when we shall have progressed farther, but we will then learn that the classification is more for convenience than because of any absolutely special properties of either class.

Many of the elements are quite rare, and are seldom seen even by chemists; others are abundant and common. The names of all will be found in the table on page 44.

The names of the non-metals which we must study are as follows:

Hydrogen.	Oxygen.	Nitrogen.	Carbon.
————	Sulphur.	Phosphorus.	Silicon.
Chlorine.	Selenium.	Arsenic.	
Bromine.	Tellurium.	Antimony.	
Iodine.		Vanadium.	
Fluorine.		Niobium.	
		Tantalum.	
		Boron.	

Because they resemble one another in important properties, certain of these elements are classed together in natural families. We will be better able to understand these relations when we have studied some of the compounds, and have seen how all of the chemical changes through which these compounds may pass can be explained by our theory of atoms and molecules. At the same time we will find that our study will greatly enlarge and render more definite the ideas which we have already acquired.

LESSON II.

HYDROGEN.

17. As we have already seen, hydrogen is one of the elements of water, of which it constitutes one-ninth by weight. It exists in combination with other elements in all animal and vegetable substances, in coal, and in the natural oils, petroleum and pitch.

In our first experiment (§ 3) we have studied one method by which it may be obtained,—the action of the metal sodium on water. That method is unsuitable for the preparation of any but very small quantities of hydrogen; when it is desired to prepare the gas from water, steam is passed over red-hot iron or zinc. The metal then combines with the oxygen of the water, setting free the hydrogen.

18. **Preparation.**—In the laboratory, hydrogen is made by the reaction of zinc with hydrochloric acid or sulphuric acid diluted with water.

We put in the bottom of a tall jar (Fig. 6) some small pieces of very thin sheet zinc, or a handful of granulated zinc, and on this pour some hydrochloric acid. A brisk effervescence begins; when we apply a flame at the mouth of the jar, the gas which is disengaged at once takes fire, and a large stream of very pale flame shoots into the air.

FIG. 6.

This gas is hydrogen. Hydrochloric acid is a compound of chlorine and hydrogen; when it acts on zinc, that metal drives the hydrogen out of its combination, and unites with the chlorine, forming a new compound, called zinc chloride. We may express the chemical change as follows:

BEFORE THE REACTION.		AFTER THE REACTION.
Hydrochloric acid containing Hydrogen + Chlorine	+ Zinc =	Zinc chloride + Hydrogen containing Zinc + Chlorine

Dilute sulphuric acid is usually employed instead of hydrochloric acid for the preparation of hydrogen. We may explain the change in a similar manner:

| Sulphuric acid containing Sulphur + Oxygen + Hydrogen | + Zinc = | Zinc sulphate containing Sulphur + Oxygen + Zinc | + Hydrogen |

It is sufficient to put the zinc in a bottle, and, after pouring in the dilute sulphuric acid, to close the mouth of the bottle with a cork through which passes a tube for the exit of the gas ; but it is more convenient to have a cork with two holes, or a bottle with two necks.   Into such a bottle (Fig. 7) we will introduce some

granulated zinc that has been made by melting zinc and pouring it from a little height into a bucket of water.   Then we adapt to one of the necks of the bottle a cork through which passes a long tube with a funnel at the upper end ; the lower end of this tube must pass nearly to the bottom of the bottle, so that it may dip into the liquid and no gas may escape by it.   To the other neck we adapt a cork bearing a tube bent at right angles, and this serves for the passage of the gas. Over the end of this tube we may pass a rubber pipe and lead the gas wherever we wish it.   We now pour through the funnel-tube some sulphuric acid which we have diluted with about five times its volume of water and allowed to cool, for sulphuric acid becomes very hot when it is mixed with water, and we always make the mixture by pouring the acid into the water, and not the water into the acid.   The effervescence shows us that gas is being disengaged, and, after waiting a few moments to allow the hydrogen time to drive all the air out of the bottle, we may make some experiments with our gas.   These experiments will make us acquainted with its properties.

Fig. 7.

19. **Properties of Hydrogen.**—Hydrogen is a colorless gas, and has neither taste nor odor, as we can determine by examining it as it escapes from the tube.   It is the lightest substance known. We connect our gas-generating bottle with the rubber pipe, the other end of which is passed over a straight glass tube, and push this tube up to the bottom of a wide test-tube which is turned upside down (Fig. 8).   In a short time this little jar is filled with hydrogen, for the gas is so light that it collects in the jar, and pushes the air down and out at the mouth.   We can prove that the jar is filled with hydrogen, for when we withdraw the tube and introduce a lighted taper, the gas at once takes fire and burns

at the mouth of the jar; the taper is extinguished on entering the gas, but is relighted as it is drawn out through the hydrogen flame. The hydrogen is collected in this case by upward *dry displacement:* it displaces the air. We again fill our tube with hydrogen in the same manner, and taking another and smaller tube we place it alongside of the first, which we carefully incline (Fig. 9) more and more until we have poured all of the hydrogen up into the second jar. On introducing a lighted taper into the latter, the gas takes fire and burns with a slight explosion, for while flowing out of the first vessel it became mixed with a little air.

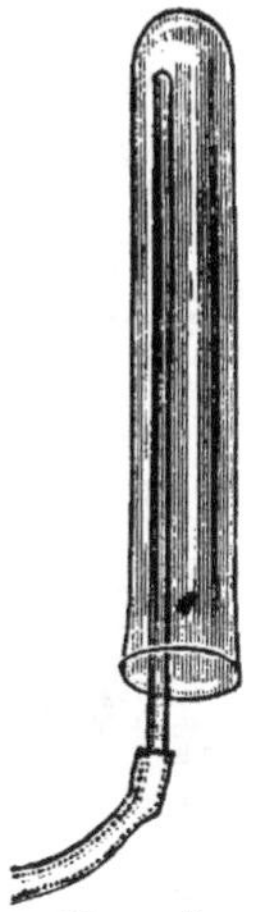

On account of its lightness, hydrogen is often used to fill balloons; soap-bubbles, which may be easily made by dipping the end of the tube into suds, will rise quickly in the air when they are shaken from the tube.

Fig. 8.

A given volume of hydrogen is only 0.0693 as heavy as the same volume of air : this is expressed by saying that the density of hydrogen compared to air is 0.0693; for equal volumes, air is then 14.44 times as heavy as hydrogen. One litre of hydrogen measured at 0° (the freezing point of water), and under the ordinary pressure of the atmosphere, weighs 0.0895 of a gramme.

20. The diffusibility of a gas is its tendency to mix with other gases: gases will mix with one another in this manner even through the pores of many substances which are sensibly porous, that is, possess

Fig. 9.

pores large enough to be seen by the aid of a microscope. It has been found that the diffusibility of gases depends upon their densities. The lighter a gas is, the more diffusible is it also, and, on the contrary, the heavier gases do not diffuse as quickly as the

lighter ones.   Since hydrogen is the lightest gas, we can understand that it must be the most diffusible: we allow a little hydrogen to escape into the air, and in a few seconds it scatters through
all the air in the room.

We have arranged another tube through which the hydrogen
may escape from our gas-bottle, and this tube is drawn out so that

it has a small opening at which we may
burn the gas if we desire.   Close above
the unlighted gas escaping from this jet
we hold a piece of paper (Fig. 10); the
hydrogen passes through the paper, as we
prove by igniting it above, and the flame
of the gas quickly sets fire to the paper
and passes through to the gas at the jet.

FIG. 10.

Because it is so diffusible, hydrogen
cannot be kept in bottles which have the
smallest cracks.   It even passes through hot plates of iron and
platinum.

21. Hydrogen is not soluble in water, and it may be collected
and kept over the pneumatic trough.   Gases which do not dissolve
in water may be collected in this manner: the jar in which we

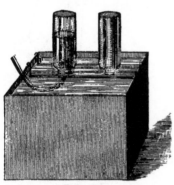

wish to receive the gas is filled
with water and inverted in a
trough near the top of which is
a shelf on which the jar may rest.
The water will not run out of the
jar as long as the mouth of the
latter is below the surface.   Under
the edge of this jar filled with
water we pass the end of the tube
from which escapes the gas that
we wish to collect; this gas bub-

FIG. 11.

bles up through the water, which it drives out of the jar (Fig.
11).   If it be desired, we can transfer the gas from one jar to
another, by first filling the second jar with water, placing it on
the trough, and then pouring the gas up through the water by

inclining the jar which contains it under the edge of that which is to receive it.

22. Hydrogen is the only gas which conducts heat.[*] We have fitted to the ends of a glass tube (Fig. 12) two tightly-fitting corks through each of which passes a smaller tube, and also a wire which may be connected with an electrical battery ; the two wires are joined by

FIG. 12.

a thin platinum wire which becomes heated red-hot by the electric current. We can now pass any gas through this tube and notice the effect on the wire: we try oxygen, nitrogen, carbon dioxide, and see that the wire still remains red-hot ; but when we pass hydrogen through the tube, the wire is cooled below redness. The hydrogen has conducted away the heat. On account of its conducting power, and because it has the property of combining with certain metals, we believe that hydrogen is a sort of metallic vapor. By very great pressure and extreme cold, hydrogen has been converted into a liquid, and even into a solid.

23. We have seen (§ 19) that hydrogen will burn in the air, and that it will not support combustion. When it burns, it combines with the oxygen of the air, forming vapor of water ; this is the sole product of the combustion of hydrogen. We may assure ourselves of this by holding over a jet of burning hydrogen a jar or any glass vessel, of which the interior will rapidly become covered with little drops of dew, and these will soon unite together and trickle down the sides of the jar. This takes place with hydrogen which has been perfectly dried by passing it through a tube containing calcium chloride, or pumice-stone wet with sulphuric acid (Fig. 13) ; both of these substances remove all moisture from gases with which they come in contact.

If hydrogen be mixed with half its volume of oxygen, or about three times its volume of air, the mixture will explode violently when ignited. For this reason we must be careful that all of the air has been driven from the generating bottle before lighting the hydrogen. We may make the explosion harmlessly by passing a little hydrogen into a hydrogen pistol, made of sheet tin, and, after

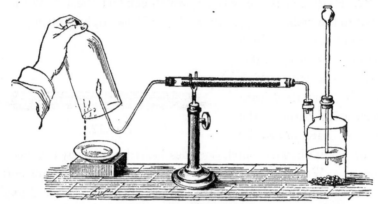

FIG. 13.

corking the mouth of the pistol, ignite the mixed gases by holding a flame to a little hole at the other end; the cork is then driven out with a loud noise (Fig. 14): we must cover the hole with the finger while we are charging the pistol.

FIG. 14.

If we slip over a small jet of burning hydrogen a rather wide glass tube (Fig. 15), we will find that when the flame has reached a certain point in the wide tube it begins to quiver, and a more or less musical tone is produced. The tone may be varied by using tubes of different lengths: it is caused by the current of air ascending the tube.

24. Certain very finely divided metals have the power of absorbing hydrogen so rapidly as to become hot enough to light the gas. Spongy platinum is such a substance; when a small piece of this very porous form of platinum, tied by a thin wire in the centre of a small brass ring (Fig. 16), is held in a jet of escaping hydrogen, it becomes bright hot and the gas is inflamed. The spongy platinum should be heated shortly before

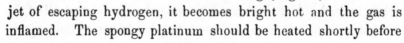

FIG. 15.

making the experiment. It is not hard to understand this phenomenon, for when the platinum absorbs
the hydrogen the gas is necessarily reduced to a small volume in the pores of
the metal, and the heat which keeps the
molecules of the gas at large distances
from each other must raise the temperature when those distances are diminished
by the condensation ; just as the heat
which converts water into steam reappears when the steam is condensed.

FIG. 16.

While hydrogen combines with many of the other elements,
the combination does not take place directly. We have seen that
heat is required to bring about the union of hydrogen and oxygen.
Hydrogen and chlorine combine under the influence of light (see
§ 73). Pure hydrogen is not poisonous, but it cannot support
respiration (see § 33).

LESSON III.

OXYGEN.—COMBUSTION.

25. Oxygen was discovered by Priestley in 1774. It is the
most abundant element at the surface of the earth : it forms about
one-fifth of the atmosphere, in which it exists uncombined, but
mixed with the element nitrogen ; its combination with hydrogen is
water, and it enters into the composition of nearly all minerals and
rocks.

We have seen (§ 5) that oxygen is produced when mercuric
oxide is heated ; but this method would be too expensive for the
preparation of large quantities of oxygen.

26. **Preparation.**—The most convenient process for obtaining
oxygen consists in heating a compound of chlorine, oxygen, and
potassium, called potassium chlorate. This is a white, crystalline

substance, from which heat drives out all of the oxygen, leaving a compound of potassium and chlorine, called potassium chloride.

We put a little potassium chlorate in a test-tube, and heat it rather strongly in the flame of a spirit-lamp or Bunsen burner.

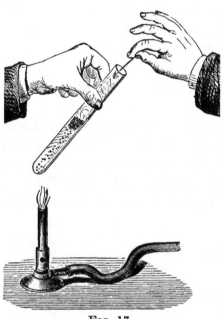

It melts, and soon begins to boil; this boiling is the escape of the oxygen, as we can prove by pushing into the tube a match-stick bearing a spark of fire, which instantly bursts into flame (Fig. 17). The white substance which remains in the tube after all of the oxygen is driven out, is the potassium chloride.

When we wish to make and collect larger quantities of oxygen, we mix the potassium chlorate with about one-eighth its weight of manganese dioxide, which causes the gas to be given off at a lower temperature, and with less danger of explosion. The manganese dioxide is a black powder, and is found unaltered after the experiment, being simply mixed with the potassium chloride : it helps the reaction because it has an affinity for oxygen ; but this affinity is so feeble that the new compound which is formed is at once decomposed, the oxygen being given off, while the manganese dioxide remains as it was at first. We may consider that it pulls the oxygen away from the potassium chlorate.

We introduce our mixture of potassium chlorate and manganese dioxide into a glass flask, to which we adapt a tightly-fitting cork bearing a tube for the exit of the gas. Then we place our flask on some dry sand in a little tin or sheet-iron dish, which we call a sand-bath, and under this we place our lamp. The sand becomes

FIG. 17.

hot and heats the flask gradually, and generally prevents cracking of the glass. We may now slip a rubber tube over the delivery-

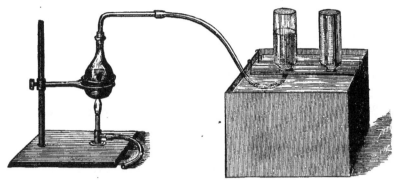

FIG. 18.

tube of the flask, and when oxygen begins to come off, as we may ascertain by holding a lighted match near the end of the tube, we may collect the gas in a jar over the pneumatic trough (Fig. 18).

As glass flasks often break in this experiment, when we want many litres of oxygen, we heat the generating mixture in a flask made of sheet copper or tin plate; to prevent leaking in a tin retort, a little white lead is put in the seams before they are lapped. As little particles of manganese dioxide are

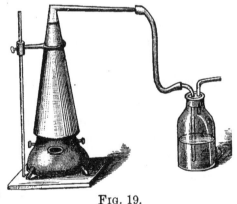

FIG. 19.

carried out with the gas, we usually wash the latter by making it pass through some water in a wash-bottle. The whole apparatus is shown in Fig. 19.

When all of the oxygen has been disengaged, we remove the end of the tube from the water in the trough before taking the heat from under the flask : otherwise water would be drawn back as the retort cools, and would break a glass flask, and the steam might

B 3

burst one of tin or copper. This precaution is observed in the preparation of all gases made by the aid of heat.

After filling several jars with oxygen, we remove them from the trough by passing a saucer under the mouth of each, below the surface of the water; then on lifting them out, the water in the saucer prevents the escape of the gas. We can now turn them quickly mouth upward, still keeping covered with the saucer, and we are ready to study the gas.

27. Properties.—Oxygen has neither color, taste, nor odor. When freshly made from potassium chlorate it usually has a smoky appearance and more or less odor, but these are impurities, and disappear after the gas has stood for a time over the pneumatic trough. It is a little heavier than the air, its density being 1.1056; one litre of the gas at $0°$, and normal pressure, weighs 1.437 grammes. It is almost insoluble in water. It has been converted into a liquid by great cold and pressure.

Oxygen manifests energetic affinity for most of the other elements: it combines with some of them at ordinary temperatures, and with others by the aid of more or less heat.

28. Combustion.—The burning of wood, coal, illuminating gas, oil, and other substances with which we are familiar, is only the combination of those bodies with the oxygen of the air. Into a small tube closed at one end, we have put some ferrous oxalate, made by adding oxalic acid to a solution of ferrous sulphate, and after drawing out the open end of the tube so as to leave a small thin opening, we twisted a wire about the tube, and heated it until no more gas escaped at the opening; we then sealed the thin

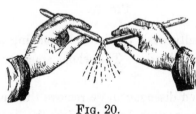

FIG. 20.

end of the tube by holding it for a moment in the flame. The result of this heating has been to decompose the ferrous oxalate, leaving a very fine powder of iron in the tube. We now break this tube, and shake out the powder, which instantly takes fire, falling in a shower of sparks (Fig. 20), if our tube has been well prepared. The iron has combined with the oxygen of the air: it has been burned into a substance called iron oxide. Usually it is necessary to heat a combustible substance before it will burn; then as soon as the union with oxygen, or the oxidation as we call it, begins, the chemical action develops sufficient heat to keep the temperature so high that the combustion may continue without further aid.

Only one-fifth of the air is oxygen, and we shall learn that the other gas with which that oxygen is mixed not only does not help, but prevents combustion: pure oxygen should then support combustion much more energetically than air, and we have seen that oxygen causes a spark on a match-stick or a taper to burst into flame.

29. We wrap a copper wire around a piece of charcoal, and fasten the other end of the wire in a hole in a piece of tin plate large enough to cover the mouth of one of our jars. After holding the charcoal in a flame until a corner of it becomes red-hot, we quickly remove the saucer from the jar and plunge into it the charcoal, which remains suspended (Fig. 21). Instantly the combustion grows very vivid, and, if we have a knotty piece of charcoal, brilliant sparks are thrown off. The charcoal combines with the oxygen until all of one or the other is used up. The result of the combination is a gas called carbon dioxide, and if we put a lighted taper into the jar containing it, the flame will be extinguished: we may say that the oxygen has also been burned, and can serve for no other combustion as long as it remains combined with the carbon.

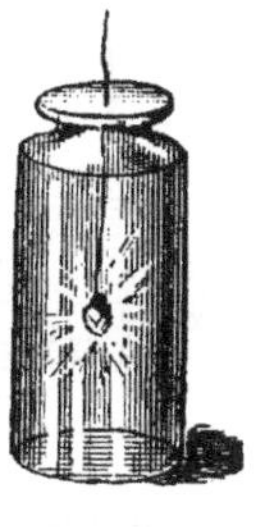

FIG. 21.

After softening a steel watch-spring by heating it in a flame, we twist it into a coil, one end of which we fasten in a cork, and over the other end we slip the split end of a piece of match-stick; after lighting this we quickly introduce it into a bottle of oxygen (Fig. 22). The flame heats the iron so hot that it can begin to burn, and the oxidation furnishes heat enough for the combustion to continue: brilliant stars of burning steel shoot out, and hot drops of iron oxide fall to the bottom of the jar, in which it is well to leave a layer of water to prevent breaking.

We have prepared a deflagrating-spoon by fastening a small, saucer-like piece of sheet copper on the end of a straight copper wire.

We support this in the hole in our tin plate, and on a little dry sand which we put in the spoon we place a piece of phosphorus (§ 177) a little larger than a pea. We light the phosphorus—it takes fire very easily—and plunge the spoon into a new jar of oxygen (Fig. 23). At once a most intense light is produced by the combustion of the phosphorus, and the jar becomes filled with a white smoke of phosphoric oxide, the compound of phosphorus and oxygen. After a time this smoke dissolves in the layer of water, which we leave in the jar for this experiment as for the last.

Fig. 22.

The metal magnesium burns very brilliantly in the air: we twist together half a dozen ribbons of this metal, and, after fastening one end in our jar-cover, we light the other with a match. On introducing this into a jar of oxygen the intensity of the combustion is dazzling. A white smoke of magnesium oxide soon settles in the jar, and contains of course the magnesium and oxygen which have combined together. The jar is usually broken in this experiment.

Fig. 23.

These experiments have been only intense cases of what we commonly call combustion, a phenomenon which we apply for the production of heat and light. The combustible substances ordinarily employed, such as wood, coal, illuminating gas, wax, tallow, oil, etc., contain carbon and hydrogen; charcoal is almost wholly carbon; these substances burn because the carbon and hydrogen which they contain, unite readily with the oxygen of the air when the union is started by the aid of heat.

30. The brightness of the light is not always proportional to

the amount of heat : we have seen that the flame of hydrogen is very pale, but it is very hot. If we desire to increase the heat of a fire, we furnish the combustible with more oxygen by blowing air into it with a bellows, and we rake the ashes from our coals in order that the oxygen may come in contact with the hot carbon : it is possible, however, to furnish too much air, if the latter be cold, as we see when we extinguish a candle-flame by blowing on it. When we want the most intense combustion possible, we supply the burning body with pure oxygen, and the hottest flame which we can obtain is that of hydrogen burning in oxygen. This flame is that of what is called the oxyhydrogen blow-pipe, in which a tube through which the oxygen is forced passes inside of another tube carrying the hydrogen ; the two gases, coming from separate gas-holders, or caoutchouc bags, mix at the opening of the jet (Fig. 24). If they were mixed before the moment of burning, the ap-

FIG. 24.

paratús containing the mixture would be burst by the explosive union of the gases (§ 23). In using the oxyhydrogen blow-pipe, we first turn on the hydrogen, light it, and then slowly turn on the oxygen until we have the hottest flame. If it be inconvenient to use hydrogen, we may substitute for it illuminating gas, connecting by a rubber tube the oxyhydrogen blow-pipe with a gas-fixture. While the oxyhydrogen flame is not very bright, it is very hot, and when we hold in it a piece of watch-spring, or an old penknife-blade, the iron is burned, making a brilliant fountain of fire. The metal platinum, which does not melt at the highest furnace heat, melts readily in the oxyhydrogen flame.

31. Fire is the combustion with incandescence—that is, production of light and heat at the same time—of a solid substance :

we have seen such phenomena in the combustion of charcoal and iron. The oxidation takes place only on the surface of the burning body.

32. Flame is the combustion with incandescence of a gas or vapor, as in the burning of hydrogen, phosphorus, and magnesium : at the borders of the flame the gas or vapor may mix with the air ; but the interior of the flame must consist of highly-heated, yet unburned gas. Why are certain flames very bright, while others give little or no light? We burn a little sulphur in a deflagrating-spoon in a jar of oxygen, and the combustion, though very brilliant, would not serve for illumination. In order to produce a bright white light, a flame must contain particles of solid matter which may become highly heated. The burning phosphorus and magnesium were brilliantly luminous because the little solid particles of phosphoric oxide and magnesium oxide which were formed, became very hot. The flames of hydrogen and sulphur do not give white light because the products of combustion, water in one case and sulphurous oxide in the other, are gases at the high temperature at which they are formed, and gases cannot be heated hot enough to give white light. However, the products of combustion of tallow, wax, and illuminating gas are not solid, yet these substances are useful for artificial light. In these cases the illumination is due to little particles of carbon. The combustible gases and vapors come in contact with enough oxygen to completely burn them only on the outer edge of the flame ; but the heat is radiated into the flame as well as from it, and the gases, which are compounds of hydrogen and carbon, are decomposed ; little solid particles of carbon are set free, and these become very hot and give out light : when they reach the outside of the flame they are entirely consumed, unless there be too little oxygen, and in that case the flame smokes. When a cold body— a piece of glass will answer—is held for a moment in the brightest part of a lamp- or gas-flame, the little particles of carbon are deposited on the cold surface in the form of soot.

We may by a very simple means render the colorless flame of hydrogen quite brilliant : we have fitted to a bottle a cork through which pass two tubes, the outer ends being drawn out to fine jets. One of these tubes is short, and passes only through the cork; the other passes to the bottom of the bottle, in which we have placed some broken pumice-stone saturated with benzine (Fig. 25).

To this same tube is joined a short side-tube, which we connect with a bottle containing zinc and dilute sulphuric acid. When all of the air is expelled from the bottle, we light the hydrogen at the two jets. At one it burns with a colorless flame: it is the flame of hydrogen just as it comes from the generating bottle. The other flame is quite bright; the hydrogen which has passed through the benzine has become charged with the vapor of that volatile liquid, and as that vapor, containing hydrogen and carbon,

FIG. 25.

is decomposed by the heat before it burns, the carbon particles become incandescent.

When the flame of illuminating gas or of a lamp is supplied with oxygen at the inside, the particles of carbon are burned instantly and do not become hot: the flame then gives no light. This is the case in the Bunsen burner (Fig. 26), in which the force of the escaping gas draws air through holes in a tube surrounding the jet; the air and gas mix together, and all of the carbon is consumed before it can become incandescent. We then have a flame which gives great heat, but does not deposit smoke on any vessels which we may heat in it.

If we hold a piece of lime in the flame of the oxyhydrogen blow-pipe, it becomes very hot and emits a brilliant light. This constitutes the calcium or oxyhydrogen light which is used in theatres. Lime is used because it is neither burned, melted, nor changed into vapor by the intense heat.

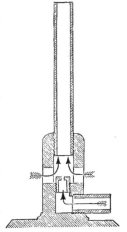

FIG. 26.

33. **Slow Combustion.**—All of the examples of oxidation which we have so far considered are said to be cases of rapid com-

bustion: they take place with the production of intense heat and light. Sometimes, however, there is no bright light, no high temperature, and yet combustion takes place as certainly as before. A piece of iron which rusts by exposure to damp air is only combining with oxygen, and the rust is a compound of iron with the oxygen and moisture of the atmosphere: here the heat of chemical union is developed so slowly that it is conducted away by the air and surrounding bodies, and the iron does not become heated.

Respiration is a slow combustion. The warmth of our bodies, and all our animal motions, are due to the gradual oxidation of the carbon and hydrogen of our tissues. At every breath fresh oxygen is introduced into the lungs, where it is absorbed by the blood and carried through the arteries to the most remote parts of the system; then, when all of the oxygen in the blood is used up, the water and carbon dioxide produced by the combustion are carried through the veins to the lungs, and thrown out with the exhaled air. Animal life itself depends on this slow oxidation: we all know how quickly any animal perishes from suffocation when the supply of air is entirely cut off. The muscles of our bodies contain no force except that which is produced by the combustion of their own substance. Great muscular exertion consequently requires increased oxidation, and we quickly become fatigued when we are obliged to burn up our tissues more rapidly than they are remade from our food. Also, the quantity and kind of food required depend upon the amount and kind of work which we must perform.

LESSON IV.

COMPOSITION OF WATER.—CHEMICAL LAWS AND THEORIES.

34. Water is the sole product of the combustion of hydrogen in air or oxygen. Its composition, that is, the proportion in which the hydrogen and oxygen are combined together, may be

determined by analysis and by synthesis. Analysis is the separation and weighing of the constituents of a compound; synthesis means the formation of a substance by causing its elements to unite in the proper proportion.

35. **Electrolysis of Water.**—Electrolysis means the decomposition of a substance by an electric current. For the decomposition of pure water an enormously strong current would be required, and because we do not desire to use such a strong current we employ dilute sulphuric acid: the final result is the same as if we were to use water, the sulphuric acid being found unchanged after the experiment. Instead of using the acid, we might make a strong solution of ordinary salt; the salt would make the water a better conductor of electricity. We introduce the dilute sulphuric acid (about five parts of water to one of acid) into a vessel through a hole in the bottom of which are cemented two wires, the inner ends of each being soldered to a little plate of thin platinum. We fill two small test-tubes with water, and, closing the mouths with the fingers, invert one over each of these plates: we now connect the ends of the wires with the poles of a voltaic battery (Fig. 27). Little bubbles of gas at once begin to rise in the tubes, and as soon as the quantities of gas collected are large enough to allow us to notice the volume of each, we see that in one of the tubes there is twice as much as in the other. When that tube is filled, we raise it carefully, and the introduction of a lighted match will convince us that the gas is

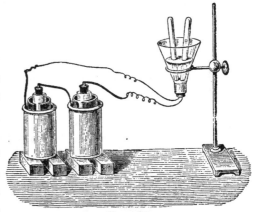

FIG. 27.

hydrogen. When we raise the other tube, keeping the end closed with the thumb, until we are ready to push into it a match-stick bearing a spark, the kindling of the spark into flame shows us that the second gas is oxygen. Water is then composed of two volumes of hydrogen combined with one volume of oxygen.

36. We have seen that the density of hydrogen compared to air is 0.0693, and that the density of oxygen is 1.1056. A given volume of oxygen must then be

$$1.1056 \div 0.0693 = 16 \text{ (a very little less)},$$

sixteen times as heavy as an equal volume of hydrogen. As we have two volumes of hydrogen and only one of oxygen, the oxygen in water must weigh eight times as much as the hydrogen.

37. **Eudiometric Synthesis of Water.**—A eudiometer is a graduated strong glass tube, closed at one end near which two thin platinum wires are soldered into the glass on opposite sides; an electric spark may be passed between the wires on the inside of the tube (Fig. 28). If we fill such a tube with mercury, and, after inverting it in a vessel of mercury, pass into it some hydrogen, and then half as much oxygen, an electric spark will cause the gases to combine; after the little explosion which takes place in the tube, the water which is formed is condensed in minute drops in the cold tube, and the atmospheric pressure forces the mercury up, filling the tube completely. Here, again, we see that water is composed of two volumes of hydrogen combined with one volume of oxygen.

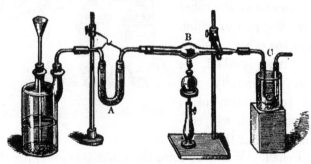

Fig. 28   Fig. 29.

38. **Synthesis by Weight.**—We may make the synthesis of

water by a very instructive method which was first adopted by the French chemist Dumas. We prepare hydrogen from sulphuric acid and zinc in the ordinary manner, and thoroughly dry it by passage through a tube (A) containing little pieces of pumice-stone wet with strong sulphuric acid (Fig. 29). We then cause it to pass through a tube containing some cupric oxide (B), a black compound of copper and oxygen, and this tube is connected with a U-shaped tube (C) filled with pumice-stone also moistened with strong sulphuric acid. The U tube is placed in a vessel containing some broken ice. Before connécting our tubes together, we have carefully weighed that holding the cupric oxide, and the last U tube with its contents. When this whole apparatus is filled with the hydrogen coming from the bottle, we heat the cupric oxide by a spirit-lamp, and when it becomes hot the hydrogen gas takes away the oxygen from the copper. Steam is formed and is condensed in the U tube (C). When the color of the cupric oxide has entirely changed to red, we warm the whole length of the tube containing it, in order to drive all of the water over into the U tube: we allow our apparatus to cool, take it apart, and again weigh the tubes of which we had determined the weight before the experiment. The copper which is left in the first will weigh just as much less than the cupric oxide as the latter has lost oxygen. The increased weight of the U tube (C) will be the weight of the water formed, and by subtracting from this weight the weight of the oxygen, we will have the weight of the hydrogen contained in that water. We find that there is almost exactly eight times as much oxygen as hydrogen. In very elaborate experiments we would perfectly purify our hydrogen and adopt all possible precautions that no vapor of water might escape from the tube C.

CHEMICAL LAWS AND THEORIES.

39. No matter by what process water may be formed, no matter by what process its composition may be determined, it is always found to contain the same proportions of oxygen and hydrogen; never more nor less than eight (7.98 exactly) parts of the first

for one óf the second. If we try to combine the gases in other proportions, the excess of the one or other, out of the proportion one to eight, will be left uncombined. The analysis of all known substances has shown a similar constancy of composition, a constancy which is expressed in the following

40. LAW OF DEFINITE PROPORTIONS: The proportion in which the elements exist in any compound, is invariable. This is generally called Dalton's first law.

41. We have already found that the proportions by volume according to which oxygen and hydrogen unite are one to two. This is a simple relation of volumes. Experiments with other gases will in time show us that when gases combine, there is always some such simple relation between the volumes of the gases that enter into combination. Thus,

One volume of hydrogen combines with exactly one volume of chlorine.
Two volumes of hydrogen combine with exactly one volume of oxygen.
Three volumes of hydrogen combine with exactly one volume of nitrogen.
Two volumes of nitrogen combine with exactly one volume of oxygen.

We might find many more such examples, and the statement of these facts constitutes

GAY-LUSSAC'S FIRST LAW: there is a simple relation between the volumes of gases which combine.

42. Let us study the volume of the compound formed when that compound is in the same condition as the original elements; that is, the gaseous state.

FIG. 30.

By grinding together with emery, we have accurately fitted together the necks of two glass bottles that have exactly the same capacity (Fig. 30). We fill the lower one with perfectly dry chlorine gas (§ 71), and the upper with dry hydrogen, and then hermetically join them together by the ground joint. All of this must be done in a room lighted only by a candle or small gas-flame. We now allow the apparatus to stand for a day in a room where the sunlight may not shine on it directly. The gases will slowly combine, and the yellowish color of the chlorine will disappear. When we open the bottles under the surface of mercury, we will find that no gas escapes from them, and no mercury enters. The gas hydrochloric acid has been formed, and

Two volumes of hydrochloric acid must contain $\begin{cases} \text{one volume of hydrogen and} \\ \text{one volume of chlorine.} \end{cases}$

Over the open end of a eudiometer we have passed a piece of strong rubber tubing, to the other end of which is attached a straight glass tube about the size and length of the eudiometer; the rubber tube is firmly tied at each joint. We fill our apparatus with mercury, place it vertically in the mercury trough, and introduce five cubic centimetres of oxygen and ten cubic centimetres of hydrogen. We now close the end of the tube with the finger, lift it from the trough, and, after bringing the two tubes parallel to each other, clamp them in that position in a stand (Fig. 31). We have tightly fitted a perforated cork around the lower end of the eudiometer, and on this cork we now as tightly fit the lower end of a wide glass tube, which we slip over the eudiometer, whose wires we have connected with long copper wires that pass out at the upper end of the wide tube. We now fill the wide tube with perfectly clear oil (lard oil or sweet oil) heated to 130°; we adjust the mercury at the same level in the two tubes, and, after carefully reading the vol-

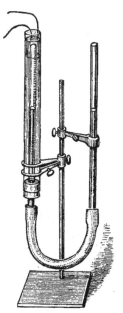

Fig. 31.

ume occupied by the mixed gases, we pass an electric spark in the eudiometer. The gases of course combine: steam is formed, but does not condense, because the eudiometer is heated. On again making the mercury levels the same, and examining the volume of this steam, we find that it is only two-thirds as great as that of the mixed gases: in other words,

Two volumes of steam contain $\begin{cases} \text{two volumes of hydrogen and} \\ \text{one volume of oxygen.} \end{cases}$

4

Ammonia gas is a compound of hydrogen and nitrogen, and its analysis proves that two volumes of the gas may be decomposed into one volume of nitrogen and three volumes of hydrogen.

On comparing these results, we find that

Two volumes of hydrochloric acid contain one volume of hydrogen and one volume of chlorine.

Two volumes of vapor of water contain two volumes of hydrogen and one volume of oxygen.

Two volumes of ammonia contain three volumes of hydrogen and one volume of nitrogen.

We see that not only is there a simple relation between the volumes of gases which combine, but, as is expressed in

GAY-LUSSAC'S SECOND LAW, there is a simple relation between the volume of a compound gas and the sum of the volumes of the gases which form that compound.

LESSON V.

CHEMICAL LAWS AND THEORIES (Continued).

43. Equivalent Combining Proportions.—Careful analysis of hydrochloric acid has shown that it is composed of 35.5 parts by weight of chlorine combined with one part by weight of hydrogen. Chlorine combines with mercury, forming a compound called mercuric chloride or corrosive sublimate; this compound contains for every 35.5 parts of chlorine, exactly 100 parts of mercury. We dissolve 135.5 grammes of mercuric chloride in water, and put some zinc into the solution: the chlorine has a stronger affinity for the zinc than for the mercury; it consequently combines with the zinc, forming zinc chloride, which remains in the solution, while mercury separates. If we wait until all of the 35.5 grammes of chlorine which were combined with 100 grammes of mercury have united with the zinc, and then determine the quantity of zinc which is required to combine with that quantity of chlorine, we would find that the zinc chloride formed weighs 68.25 grammes: that is, 32.75 (68.25 — 35.5) grammes of zinc combine with 35.5 grammes of chlorine. Consequently, as far as combining with chlorine is concerned, 32.75 parts of zinc have just as much power as 100 parts of mercury.

Oxygen combines with mercury and with hydrogen: it combines also with zinc and with chlorine. Analysis of the compouuds so formed shows that 35.5 parts of chlorine will combine with 8 parts of oxygen, and that 8 parts of oxygen will combine with 32.75 parts of zinc or with 100 of mercury. We have already seen (§ 36) that 8 parts of oxygen combine with one part of hydrogen. These numbers must then express the relations between the combining quantities of the corresponding elements. Thousands of analyses have shown that similar equivalent proportions exist for all of the elements. The combining proportions so found must bear simple relations to the relative weights of the atoms; for if atoms have a real existence, chemical combination must result from the union of one, two, or more atoms of one element with one, two, or more atoms of another; and, since combination is in definite proportions, the same substance must always result from the union of the same kind of atoms in the same proportion.

44. If it be possible to determine the relative weights of the molecules, compared with any unit, and to arrive at definite conclusions as to the number of atoms these molecules contain, then we can determine the relative weights of the atoms. Some new considerations will enable us to make such determinations.

LAW OF AVOGADRO AND AMPÈRE.—ATOMIC THEORY.

45. Different solid and liquid substances expand in very different degrees by the action of the same temperature. Gases, however, all expand alike. If at the same pressure we raise or lower the temperature of equal volumes of different gases through the same number of degrees, we find that they all expand or contract precisely the same proportion of the volume. If expansion be separation of molecules from one another (§ 8), it follows that *equal volumes of gases, measured at the same temperature and*

pressure, contain the same number of molecules. This hypothesis, proposed by Avogadro and Ampère, is true if there be such things as molecules, and if there be no molecules we can explain no chemical phenomena.

46. If equal volumes of gases contain the same number of molecules, the relative weights of those equal volumes must also express the relative weights of the molecules. The relative weights are the densities, and these densities are usually calculated to express the relation between the weight of the gas and that of an equal volume of air, which is taken as unity. Since hydrogen is the lightest gas, its molecule must have the least weight: the density of oxygen is sixteen times as great as that of hydrogen, and the molecule of oxygen must be sixteen times as heavy as that of hydrogen. Because of the lightness of the molecule of hydrogen, chemists have chosen that molecule as the standard of comparison for other molecules, and consequently the unit of density. The density of hydrogen compared to air being 0.0693, the air is 14.44 times as heavy as hydrogen: consequently if we know the density of a gas compared to air, we may easily calculate its density compared to hydrogen by multiplying the first by 14.44. Thus, the density of vapor of water compared to air is 0.622, compared to hydrogen it is $0.622 \times 14.44 = 9$. The molecule of water (steam) must then be nine times as heavy as that of hydrogen.

47. Let us see whether we can determine the relations between the weight of any atom and that of a molecule of hydrogen; and for this we must refer to the experiments which led us to Gay-Lussac's laws (§ 42). According to the law of Avogadro and Ampère, the molecule of oxygen is sixteen times as heavy as that of hydrogen. Hydrogen combines with only half its volume of oxygen, and if water consist of one atom of hydrogen combined with one atom of oxygen, the last must be eight times as heavy as the first: this would make the molecule of water nine times as heavy as the atom (or molecule) of hydrogen. But hydrogen and chlorine combine in equal volumes, and no contraction results from the combination (§ 41). Since the chlorine is 35.5 times as heavy as the hydrogen, the molecular weight

of the compound, or its density referred to hydrogen, is $\frac{1}{2}$ (35.5 + 1) = 18.25. But this molecule would contain only half as much hydrogen as combined with the oxygen to form a molecule of water; we supposed the hydrogen in a molecule of water to be only one atom, and atoms are indivisible. Therefore we must conclude that the two volumes of hydrogen which combine with one volume of oxygen represent two atoms, and the atom of oxygen is then sixteen times as heavy as that of hydrogen.

48. Now we may apply our theory to the facts already studied.

Two volumes of hydrogen represent two atoms, each of which weighs one: two volumes of oxygen represent two atoms, each of which weighs sixteen. Water is formed by the union of two atoms of hydrogen and one atom of oxygen, and a molecule of water weighs eighteen times as much as an atom of hydrogen.

A molecule of hydrochloric acid contains one atom of hydrogen and one atom of chlorine, and this molecule weighs 36.5 if one atom of hydrogen weighs 1.

A molecule of ammonia contains one atom of nitrogen (weighing 14) and three atoms of hydrogen, and is 17 times as heavy as an atom of hydrogen.

But the density of water compared to hydrogen is 9; that of hydrochloric acid, 18.25, and that of ammonia, 8.5. We see then that if an atom of hydrogen occupies one volume and weighs one, the molecule of any gas or vapor must occupy twice as much volume as one atom of hydrogen, and the weight of the molecule will be expressed by twice the density referred to hydrogen. We must remember that the molecule of hydrogen contains two atoms, each of which weighs 1.

Different methods are employed for determining the atomic weights of the elements. At this point we need only understand that if the element be gaseous or volatile, and if we have reason to believe that its molecule contains two atoms, then, since the molecule of hydrogen consists of two atoms, and equal volumes of gases contain equal numbers of molecules, the same figures which express the density of the gas compared to hydrogen, will express also the atomic weight.

This atomic theory, which has been slowly developed during the present century, furnishes an intelligible explanation of chemical phenomena. New discoveries are continually bringing new facts to its support, and, though it may be modified by the results of future researches, its principal features will probably remain undisturbed.

CHEMICAL NOTATION.

49. In order that the composition of a substance, that is, the number and kinds of atoms in its molecules, may be understood at a glance, we employ a special method of representing elements and compounds. The first letter, or the first and another letter, of the name of an element, is used to express one atom of that element. Thus, H means one atom of hydrogen ; O, one of oxygen ; S, one of sulphur ; C, one of carbon, and Ca, one of calcium. These letters are called the symbols of the elements. More than one atom is expressed by a little figure placed to the right of the symbol, slightly above or below its central line ; H^2 (read, H two) means two atoms of hydrogen ; O^4 represents four atoms of oxygen. Compounds are then written so that the symbols entering into the formulæ express the number and kind of atoms in a molecule. H^2O means a molecule of water, composed of two atoms of hydrogen and one of oxygen : H^2SO^4 means a molecule of sulphuric acid, containing two atoms of hydrogen, one of sulphur, and four of oxygen. To express any number of molecules we use an ordinary figure placed to the left of the formula ; thus, $2HCl$ means two molecules of hydrochloric acid, each of which contains one atom of hydrogen and one of chlorine.

50. We may now study the molecular changes which have occurred in the chemical phenomena that we have already observed. In the decomposition of water by sodium, one atom of hydrogen in each molecule of water is replaced by sodium, and when a molecule of hydrogen is set free, two molecules of a compound called sodium hydrate are formed. We represent the change thus :

$$2H^2O \quad + \quad Na^2 \quad = \quad 2NaOH \quad + \quad H^2$$

| 2 molecules of water. | 2 atoms of sodium. | 2 molecules of sodium hydrate. | 1 molecule of hydrogen. |

This chemical equation expresses the changes which take place in the chemical reaction. As the symbol for each atom means a definite quantity of matter, and as there can be no change in the quantity of matter during the reaction, there must be as many atoms represented in one member of the equation as in the other.

When we know what is formed by the reaction of certain molecules, our equation will enable us to calculate the quantities of the substances. The weight of one atom of sodium being 23; one atom of hydrogen, 1; and one atom of oxygen, 16, we find that 46 grammes of sodium will yield 2 grammes of hydrogen, and 80 grammes of sodium hydrate.

$$HOH \quad + \quad Na^2 \quad = \quad 2NaOH \quad + \quad H^2$$
$$1 + 16 + 1 \quad 23 + 23 \quad 2(23 + 16 + 1) \quad 1 + 1.$$

We can calculate the volume of the hydrogen at $0°$, from its weight (§ 19), and we can so estimate the quantity of hydrogen which will be set free by any quantity of sodium.

The action of hydrochloric acid on zinc (§ 18) can be expressed

$$Zn \quad + \quad 2HCl \quad = \quad ZnCl^2 \quad + \quad H^2$$
Zinc. Hydrochloric acid. Zinc chloride. Hydrogen.

The action of sulphuric acid on zinc, which yields zinc sulphate and hydrogen, is written

$$Zn + \quad H^2SO^4 \quad = \quad ZnSO^4 \quad + H^2$$
Sulphuric acid. Zinc sulphate.

Of course we must know by experiment what is formed in a reaction, before we can write the chemical equation; we must also know by analysis the proportions of the elements in any compound before we can write a formula which we believe to express the atoms in its molecule.

The decomposition of potassium chlorate by heat (§ 26) is

$$2KClO^3 \quad = \quad 2KCl \quad + \quad 3O^2$$
Potassium chlorate, Potassium chloride, Oxygen,
2 molecules. 2 molecules. 3 molecules.

That of mercuric oxide, in the same manner, is

$$2HgO \quad = \quad 2Hg \quad + O^2$$
Mercuric oxide. Mercury.

The reaction of hydrogen with cupric oxide, which enabled us to make the synthesis of water, is written

$$CuO \quad + H^2 = H^2O + \quad Cu$$
Cupric oxide. Copper.

51. The following table gives the names and symbols of the elements which are at present known, and the weights of the atoms compared to the weight of an atom of hydrogen. Some of these atomic weights might be more exactly expressed; an atom of oxygen is 15.96 times as heavy as that of hydrogen; the exact atomic weight of nitrogen is 14.01; but these numbers are so nearly 16 and 14, that for memory's sake it is preferable to use the nearest whole numbers.

NAMES OF THE ELEMENTS.	Symbols.	Atomic Weights.	NAMES OF THE ELEMENTS.	Symbols.	Atomic Weights.
Aluminium	Al	27.5	Molybdenum . .	Mo	96
Antimony (stibium)	Sb	120	Nickel	Ni	59
Arsenic	As	75	Niobium	Nb	94
Barium	Ba	137	Nitrogen	N	14
Bismuth	Bi	210	Osmium	Os	199.2
Boron	Bo	11	Oxygen	O	16
Bromine	Br	80	Palladium . . .	Pd	106.6
Cadmium	Cd	112	Phosphorus . . .	P	31
Cæsium	Cs	133	Platinum	Pt	197.5
Calcium	Ca	40	Potassium (kalium)	K	39.1
Carbon	C	12	Rhodium	Rh	104.4
Cerium	Ce	141.2	Rubidium . . .	Rb	85.2
Chlorine	Cl	35.5	Ruthenium . . .	Ru	104.4
Chromium . . .	Cr	52.5	Samarium . . .	Sa	150
Cobalt	Co	59	Scandium . . .	Sc	44
Copper	Cu	63.5	Selenium	Se	79.5
Didymium . . .	Di	145.4	Silicon	Si	28
Erbium	Er	166	Silver (argentum) .	Ag	108
Fluorine	Fl	19	Sodium (natrium) .	Na	23
Gallium	Ga	69.	Strontium . . .	Sr	87.5
Glucinum	Gl	9.5	Sulphur	S	32
Gold (aurum) . .	Au	197	Tantalum	Ta	182
Holmium	Ho	162 (?)	Tellurium . . .	Te	128
Hydrogen . . .	H	1	Thallium	Tl	204
Indium	In	113.4	Thorium	Th	234
Iodine	I	127	Tin (stannum) . .	Sn	118
Iridium	Ir	198	Titanium	Ti	50
Iron (ferrum) . .	Fe	56	Thulium	Tu	170.4(?)
Lanthanum . . .	La	139	Tungsten (wolframium)	W	184
Lead (plumbum) .	Pb	207	Uranium	Ur	120
Lithium	Li	7	Vanadium	V	51.37
Magnesium . . .	Mg	24	Yttrium	Y	172.5
Manganese . . .	Mn	55	Zinc	Zn	65.2
Mercury (hydrargyrum) . . .	Hg	200	Zirconium . . .	Zr	90

LESSON VI.

PROPERTIES OF WATER.—POTABLE AND MINERAL WATERS.

52. Pure water is not met with in nature. When we desire it, it must be prepared by distilling water; that is, boiling it in a retort, and condensing the steam. We usually conduct this distillation in tin-lined copper retorts. For most of the distillations in the laboratory we employ a flask or retort, connected with a long tube which is surrounded by a wider tube, and a stream of cold

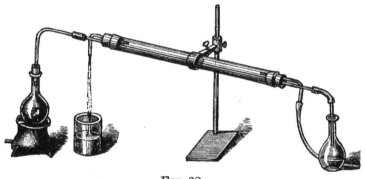

FIG. 32.

water continually flows through the space between the two tubes in this condenser, as we call it (Fig. 32).

Pure water has neither taste nor odor; although it is colorless in small quantity, it has a deep-blue color when in large masses. It solidifies when sufficiently cooled, and, since it is always converted into ice at the same temperature, that temperature is taken as 0° in the centigrade thermometer scale which we use in the laboratory.

53. The temperature of water does not change while it is freezing, and that of ice does not change while it is melting. This is because all of the heat which is communicated to ice during its melting is required to produce the change of state; indeed, one

kilogrammé of ice at 0° requires as much heat to melt it as would raise 79 kilogrammes of water from 0° to 1°, or one kilogramme from 0° to 79°, and yet the water from the melted ice still has a temperature of 0°. Ice is crystallized; it consists of a great many little six-sided pyramids dovetailed together. We can notice the

FIG. 33.

crystalline form of water by examining some snow-flakes that have fallen on black cloth (Fig. 33).

During the cooling of water, it contracts in volume until its temperature reaches 4°; it then begins to expand, and on freezing expands considerably. Ice is only 0.93 as heavy as water at 4°. Strong vessels are broken by the freezing of water in them, and it is the same expansion which kills delicate plants by frost, for the ice formed in them tears apart the fibres and destroys the sap-vessels. Since it is easy everywhere to obtain water at its point of maximum density, that is, 4°, this density has been chosen as the unit of density or specific gravity for liquids and solids. It is also at this temperature that one litre of water weighs one kilo-gramme.

Water and ice continually emit invisible vapor, but water does not begin to boil until its tension of vapor * is equal to the at-mospheric pressure. We consider that the normal atmospheric pressure is equal to 760 millimetres of mercury, and under this

---

* The tension of vapor of a liquid at any temperature is measured by the decrease in the height of the mercury in a barometer-tube, up into which the liquid is passed in small quantities until no more of it changes into vapor. The number of millimetres through which the level of the mercury has then fallen, expresses the tension of the vapor.

pressure the boiling point of water is selected as the 100° point in the centigrade thermometric scale which we use.

While water is boiling, its temperature does not rise : after it has reached the boiling point, all the heat passes into the steam, where it is required to hold apart the molecules. To convert one kilogramme of water at 100° into steam requires enough heat to raise the temperature of 537 kilogrammes of water from 0° to 1°, or 5.37 kilogrammes from 0° to 100°. The conversion of water into steam expands it 1696 times : that is, one litre of water will yield 1696 litres of steam at 100°.

54. **Chemical Properties.**—We have seen that water is decomposed by an electric current : it is also decomposed by very high temperatures (1200°). We will find that it enters into many chemical reactions, in some of which it is decomposed and part of its hydrogen set free, as in the experiment with sodium (§ 3). In other cases both the oxygen and hydrogen atoms are taken into new combinations. Water forms a large proportion of animal and vegetable tissues.

Water dissolves many substances, solid, liquid, and gaseous. We all know that salt, sugar, and alum will dissolve in water, becoming for the time part of the liquid. We immerse the bulb of a thermometer in a vessel of water, into which we throw some ammonium nitrate, and stir the liquid : at once the thermometer indicates a lower temperature. When solids dissolve in water, cold is produced, because the heat required to separate the molecules of the solid must be taken from the water. On the contrary, when gases dissolve in water, the liquid becomes warmer, because the heat no longer required to hold apart the molecules of the gas can now raise the temperature.

55. Water exerts a very curious action on some substances. We take some large blue crystals of cupric sulphate and heat them on a piece of tin over a lamp. We see that they gradually become white, and crumble into a powder. We throw some of this powder into water, and the water becomes blue ; cupric sulphate can only exist in crystals when it is combined with water, and it is blue only when combined with water. In the same

manner water is necessary to the crystalline form, and often to the color, of many substances, and when combined in this way is called water of crystallization. Water of crystallization is chemically combined in the crystals, for it is always in definite proportions. In a piece of crystallized cupric sulphate there are five molecules of water for every molecule of copper sulphate.

56. NATURAL WATER.—As it occurs naturally, water always contains foreign matters suspended or dissolved in it. These substances are derived from the air through which the rain falls, or from the soil over which the water flows.

57. According to the kind and quantity of these matters present, the water is potable, mineral, or unfit for drinking and cooking. Potable or drinking water should be cool, limpid, and odorless, having a very feeble but pleasant taste that should be neither bitter, salty, nor sweet. It should also make suds with soap without forming a curd. Water which possesses these properties always holds in solution a certain quantity of the gases of the air, oxygen, nitrogen, and carbon dioxide, and usually a small quantity of mineral matters. The gases are absolutely essential to good water, but their quantity varies considerably in different waters, and at different times in water from the same source. This dissolved air separates from water which stands in a warm place, and part of it collects on the sides of the vessel in small bubbles, which we have all seen in a glass of water that has stood for several hours. Fish cannot live in water containing no dissolved oxygen; they do not breathe, but their gills remove the dissolved oxygen from the water which they continually draw through those organs (see § 33).

The solid matters in a potable water should not exceed two or three decigrammes per litre, and these matters should be entirely mineral. They usually consist of compounds of calcium and magnesium; magnesium sulphate and calcium sulphate being the most common, while a small proportion of common salt is generally present.

58. When larger quantities of calcium compounds are present, we have no longer a soft but a *hard water*. Hard waters contain

either calcium sulphate or calcium carbonate. Water dissolves only a very small quantity of calcium sulphate, but then has a peculiar taste and curdles the soap when we use it for washing. We add a few drops of a solution of barium chloride to some water containing a little calcium sulphate, and instantly a white cloud appears : this is caused by the formation of an insoluble body called barium sulphate, and the test makes us sure that the water contained a sulphate. Calcium carbonate is insoluble in pure water, but it dissolves in water containing carbonic acid gas, or carbon dioxide. When such water is boiled, the carbon dioxide is driven out, and then the calcium carbonate separates, for it is no longer soluble. Hence we have a method of curing hard water which contains only calcium carbonate and carbonic acid : we boil it, and allow it to settle, and after pouring off the clear water expose it to the air for a time, so that it may dissolve some of the gases from the atmosphere.

59. Drinking-water must not contain animal or vegetable substances : they render it very unwholesome. Happily, the waters of rivers, which become contaminated with so many such impurities, generally become purified during their exposure to the air, because the foul matter is gradually oxidized. Water containing these matters usually has a sweetish taste and a disagreeable odor, which may, however, be very faint. It may be purified by passing it through a charcoal filter (§ 227).

60. MINERAL WATERS contain various dissolved mineral matters. Some are hot, others warm, and still others cold. Those which effervesce or sparkle contain a considerable proportion of carbonic acid gas in solution, and it is the escape of this gas which produces the sparkling. Apollinaris water contains, besides the carbonic acid gas, principally a little sodium acid carbonate, common salt, and magnesium and sodium sulphates. Buffalo lithia water contains very little of the sodium compounds, but considerable quantities of calcium sulphate, with carbonates of potassium, calcium, barium, and lithium.

Saratoga water has a large proportion of calcium and magnesium carbonates dissolved by the excess of carbon dioxide which it contains, and a very large proportion of common salt. Gettysburg water contains principally the carbonates of calcium, magnesium, and sodium, together with a little dissolved silica : its excess of carbon dioxide is quite small. Hunyádi Janos contains sulphates of magnesium, sodium, calcium, and potassium ; these substances give to it purgative properties, in which it is resembled by Friedrichshall water, for the composition of the latter is somewhat similar.

Chalybeate waters are such as contain either iron carbonate, held in solution by an excess of carbon dioxide, or ferrous sulphate : in the former case the water becomes muddy on exposure to the air, for as the carbon dioxide escapes, ferrous carbonate is deposited. The Mercer County water, of Virginia, contains a large proportion of ferrous sulphate. Iron waters are usually cold.

Sulphur waters owe their odors and their virtues to hydrogen sulphide and sulphides of potassium and sodium. They are generally warm, or even hot.

LESSON VII.

NOMENCLATURE OF COMPOUNDS OF OXYGEN.— OZONE.—HYDROGEN DIOXIDE.

61. Besides being able to express the composition of molecules by chemical formulæ, as we have learned, it is important that we may have distinctive names for each substance, and that those names may express as far as possible the composition of the molecules. A compound of only two elements is called a binary compound ; one containing three is a ternary compound ; one containing four, a quaternary. We may be satisfied at present to study a system of naming—a nomenclature—for the binary compounds of oxygen. These are called *oxides*.

We place a small piece of phosphorus on a piece of glass on a plate, light it, and cover it with a bell-jar (Fig. 34).   The phos. phorus combines with the oxygen of the air, and the compound which is formed settles like flakes of snow in the jar and on the plate.   This is an oxide of phosphorus.   On another plate, in the same manner, we burn a small piece of sodium : we have here formed sodium oxide.   Now we rinse out each jar and plate with a little water : when the water comes in contact with the oxides that have been formed, there is a hissing noise, and the jars become

FIG. 34.

warm, showing that energy has been developed ; there is a chemical action between the water and the oxide.   We pour into separate vessels the liquids from the two jars, and to that from the phosphorus oxide we add some blue litmus solution, prepared by boiling litmus, a substance made from a peculiar moss, with water. The blue color instantly changes to red.   We pour some of this red liquid into the water from the sodium experiment, and the blue color at once reappears.   It is certain then that our two oxides have different properties, and the study of these and other oxides has shown that when oxygen combines with a *non-metallic element*, the resulting oxide usually *combines with water*, and forms a substance which changes blue litmus to red.   Such substances generally have a sour taste, and are called *acids*.   On the contrary, the oxides of the metallic elements usually change the reddened litmus to blue, and are called basic oxides.   Some oxides, however, have no effect on either red or blue litmus.

62. When an oxide reacts with water, a body called a *hydrate* is formed.   The acids containing oxygen are hydrates corresponding to non-metallic oxides, while the metallic oxides usually have

corresponding *metallic hydrates.* We have seen the formation of
sodium hydrate by the action of sodium on water; this same compound results from the action of water on sodium oxide, and we
will notice that the only difference between the hydrate and water
is that the former contains an atom of sodium in place of one atom
of hydrogen.

$$Na^2O \quad + \quad H^2O \quad = \quad NaOH \quad + \quad NaOH$$
Sodium oxide.   Water.   Sodium hydrate.   Sodium hydrate.

63. An analysis of the oxide of phosphorus which we have
formed, shows that its molecule contains one atom of phosphorus
and five atoms of oxygen; it is therefore called *phosphorus
pentoxide,** and, because the acid which it forms is called phosphoric acid, the oxide is sometimes called phosphoric oxide.
In general, the name of an oxygen compound is formed by putting
oxide after the name of the other element, and to the word oxide
is prefixed the Greek name of the number of atoms of oxygen in a
molecule of the oxide.

A *monoxide* contains one atom of oxygen, a *dioxide* contains
two atoms of oxygen, a *trioxide* contains three, a *tetroxide* contains four, a *pentoxide* five.

The word *sesquioxide* is sometimes used to indicate a compound whose molecule contains three atoms of oxygen and two
atoms of the other element: *sesqui* means one and a half. Manganese sesquioxide contains Mn^2O^3.

Sometimes an element forms more than one compound with
oxygen. Nitrogen forms five; and when we have learned that a
molecule of each of these oxides contains two atoms of nitrogen,
the names will at once indicate the composition of the molecules.

Nitrogen monoxide, N^2O.
Nitrogen dioxide, N^2O^2.
Nitrogen trioxide, N^2O^3.
Nitrogen tetroxide, N^2O^4.
Nitrogen pentoxide, N^2O^5.

64. Frequently when there are only two oxides of an element,
or when there are two of special importance, the word oxide is

---

* *Penta*—five.

not changed, but the name of the other element is made to end in *ic* or *ous*.   There are only two oxides of mercury ; that containing the largest proportion of oxygen is called mercuric oxide, while that containing the least proportion is mercurous oxide.   Their molecules contain

Mercuric oxide, HgO.
Mercurous oxide, Hg^2O.

Each of the two more important oxides of sulphur contains one atom of sulphur combined respectively with three and two atoms of oxygen.

Sulphuric oxide, SO^3.
Sulphurous oxide, SO^2.

The oxide whose name ends in *ic* then contains a larger proportion of oxygen than that whose name ends in *ous*, and we should not use these terminations unless there be two oxides of the element.

We can now understand what is meant when we say that water is hydrogen oxide, and we will presently learn the signification of all the names which we have been obliged to use.

OZONE.

65. Before, and sometimes during, a thunder-storm, there is often a peculiar odor in the air, and the same odor may be noticed near a good electric machine in operation.   It has been found that the air has at the same time acquired more active oxidizing properties than it had before.   It will even bleach many coloring matters.   Part of the oxygen of the atmosphere has been changed to a body which we call ozone.

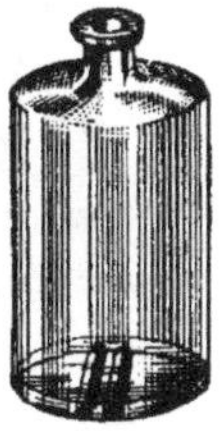

66. We can produce this change by a simple experiment.   Under the surface of some water we scrape the outside of a stick of phosphorus, so that it may be perfectly free from oxide, and then put it into a bottle containing enough water to about half cover the phosphorus, so that it may not take fire (Fig. 35).   After

FIG. 35.

it has stood for a little while, we dip into the air in the bottle a piece of paper that has been soaked in some starch boiled in water to which a little potassium iodide has been added.   We see that the paper at once becomes blue.   Now let us put a drop of a solution of iodine in alcohol on a piece of paper soaked in starch to which no potassium iodide was added.   The same blue color appears.   Potassium iodide is a compound of potassium and iodine, and the blue color is due to the action of ozone, which takes the potassium away from the iodine : as soon as the latter becomes free, it produces the blue color with the starch (§ 91).

If we suspend a bright silver coin in ozone, it soon becomes tarnished.   If we smell the air in the bottle, we find that it has a peculiar, and not very pleasant, odor.

It has been found that these same phenomena are produced by pure oxygen gas through which electrical sparks have been passed (Fig. 36), and that by the passage of such sparks the volume of the oxygen is diminished, while its weight of course does not change.   The increase in density so observed has shown that ozone is half again as heavy as oxygen : when ozone is heated, it is converted into ordinary oxygen, and the volume is expanded in the same proportion. Chemists have consequently been led to believe that while ordinary oxygen contains two atoms in its molecule, a molecule of ozone contains three such atoms.   We may consider, therefore, that if a molecule of ordinary oxygen is represented by the formula O^2, O^3 represents a molecule of ozone.

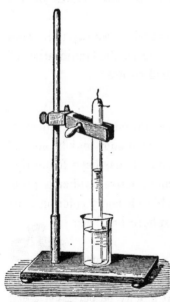

FIG. 36.

67. Let us see why ozone possesses more active powers than oxygen. When we pass electric sparks through oxygen, we decompose its molecules, and the energy of electricity is transferred to the atoms, which it enables to combine by threes, instead of by twos.   Phosphorus is gradually oxidized by oxygen, but one atom of phosphorus does not combine with whole molecules of oxygen : we

shall in time learn that in this case two atoms of phosphorus take
three atoms of oxygen; that would be a molecule and a half; but,
while the energy developed by the rapid combustion of phosphorus
appears as heat and light, the energy developed by the slow combustion of the phosphorus is transferred to the odd atom of oxygen, and enables it to combine with two other atoms set free from
molecules in the same manner. We might represent this by our
symbols.

$$6P \quad + \quad 6O^2 \quad = \quad 3P^2O^3 \quad + \quad O^3$$

Phosphorus,	Oxygen,	Phosphorus trioxide,	Ozone,
six atoms.	six molecules.	three molecules.	one molecule.

As ozone contains more energy than oxygen, we naturally
expect its properties to be more energetic.

We shall have occasion to study many actions of this kind, where
the energy evolved by the combination of certain atoms is transferred to other atoms, giving them more active properties than
they had before.

When ozone oxidizes other bodies, in most cases only one of its
atoms is used in the oxidation; the other two unite to form a
molecule of oxygen.

When the moist potassium iodide was decomposed by ozone,
potassium hydrate was formed; its molecule contains KOH, and
we see that the water present must have taken part in the reaction, which we may write

$$2KI \quad + H^2O + O^3 = \quad 2KOH \quad + O^2 + I^2$$

Potassium iodide.		Potassium hydrate.	

68. Ozone is produced in nearly all slow combustions. By
cold and pressure it has been converted into a sky-blue liquid.
It is destroyed, that is, converted into oxygen, by a temperature
of 290°. Its oxidizing powers are sometimes employed for bleaching and disinfecting, the ozone in these cases being produced by
electricity.

HYDROGEN DIOXIDE, H^2O^2.

69. We introduce some pulverized barium dioxide, a compound
whose molecule contains one atom of the metal barium and two
atoms of oxygen, into a small flask containing some cold dilute hydrochloric acid; as the solid dissolves, a solution of barium chloride

is formed, while the hydrogen of the hydrochloric acid and the oxygen of the barium dioxide combine to form a compound called hydrogen dioxide, which remains dissolved in the liquid.

$$BaO^2 \quad + \quad 2HCl \quad = \quad BaCl^2 \quad + \quad H^2O^2$$
Barium dioxide.   Hydrochloric acid.   Barium chloride.   Hydrogen dioxide.

The separation of the hydrogen dioxide from the barium chloride is not an easy matter, but the latter compound will not interfere with our experiments.

We pour some of the liquid on a little manganese dioxide; at once a brisk effervescence takes place, and by the aid of a matchstick bearing a spark, we are shown that the tube is filled with oxygen. The hydrogen dioxide has been decomposed into water and oxygen; but the manganese dioxide remains unchanged; it is probably in fact converted into a higher oxide, but the additional oxygen is at once taken away from this oxide by another atom of oxygen with which it forms a molecule of the gas.

In another tube, we pour a little of our solution on some black lead sulphide: the color quickly changes to white, and no gas is given off, for the lead sulphide is converted into lead sulphate, while water is formed.

$$PbS \quad + 4H^2O^2 = \quad PbSO^4 \quad + 4H^2O$$
Lead sulphide.   Lead sulphate.

We now pour a little hydrogen dioxide into some purple solution of potassium permanganate. The color is at once destroyed, and the liquid becomes colorless; at the same time bubbles of oxygen are disengaged, and may be identified by the usual test. In this case an atom of oxygen, very loosely held by the other atoms in the hydrogen dioxide, has combined with another atom from the potassium permanganate, which is very rich in oxygen, and a molecule of free oxygen is given off, while water is formed as before.

We mix some of the hydrogen dioxide liquid with a little yellow solution of potassium dichromate; we then quickly pour in a quantity of ether, and briskly shake the tube; the ether being lighter than the water, comes to the top of the latter, in which it is almost insoluble, and this layer of ether has a dark blue color. It contains perchromic acid, a body which is formed by the oxidation of the potassium dichromate; but with hydrogen dioxide this perchromic acid behaves just like potassium perman. ganate; unless it is at once removed from the liquid in which it is formed, its oxygen is taken away, and a green liquid containing a lower oxide of chromium is obtained.

70. We may then conclude that hydrogen dioxide acts in three ways with other substances: sometimes it is reduced, that is, part of its oxygen is taken away, while the other body remains unchanged, as is the case with manganese dioxide; sometimes the second substance is oxidized, as with the lead sulphide and potassium dichromate; sometimes both the hydrogen dioxide and the other body are deoxidized, as in the cases of potassium permanganate and perchromic acid.

Pure hydrogen dioxide is a syrupy, colorless liquid, without odor, and having a density of 1.45. It is slowly decomposed into water and oxygen at ordinary temperatures, with brisk effervescence at 100°, and explosively if dropped on a surface heated to higher temperatures.

Hydrogen dioxide and ozone undergo mutual decomposition, water and free oxygen being formed.

$$H^2O^2 \quad + \quad O^3 \quad = \quad H^2O \quad + \quad 2O^2$$

Hydrogen dioxide.   Ozone.   Water.   Two molecules of oxygen.

We must consider that a molecule of hydrogen dioxide is formed by the union of two atomic groups, each containing an atom of oxygen and an atom of hydrogen. The oxygen atoms in this molecule contain more energy than the similar atoms in molecules of water and of free oxygen, and this extra energy is manifested when the hydrogen dioxide decomposes.

LESSON VIII.

CHLORINE.—CHLORIDES.

Atomic weight, 35.5.   Symbol, Cl.

71. The element chlorine has such strong affinities for other elements that it does not exist free in nature, but is always found

in combination. The most important of its compounds is common salt, which contains sodium and chlorine, and of which enormous quantities exist in the ocean, and in salt springs and salt mines. We do not usually prepare chlorine directly from salt, but from hydrochloric acid, the latter being prepared from the salt itself.

We mix in a glass flask some strong hydrochloric acid with about one-sixth its weight of manganese dioxide, and, after adapting to the flask a cork through which passes a tube for the exit of the gas, and another tube called a safety-tube, we gently heat the mixture over a flame (Fig. 37). The safety-tube (A), which

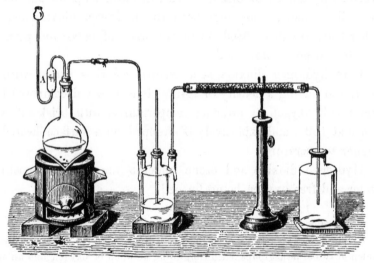

FIG. 37.

is bent around on itself and has a little bulb blown on the bend, enables us to add more acid if necessary, and at the same time if there should be too much pressure in the flask it allows the gas to escape through the little acid which we must pour into it: on the contrary, when the flask cools and the gas contracts in volume, air may enter through the safety-tube, and any liquid into which we may wish the end of the delivery-tube to dip, will not be drawn back into the flask. We may dry our chlorine gas by passing it through a bottle containing either calcium chloride or strong sulphuric acid, or we may pass it directly into a bottle.

Chlorine dissolves in water, and we collect it by downward dry displacement; for it is a heavy gas, and when we pass the tube through which it flows to the bottom of a jar, the chlorine gradually forces the air out at the top. We might collect it over salt water in the pneumatic trough, as it does not dissolve in salt water. We can easily see when the jar is full, for the gas has a greenish-yellow color. While we are filling several jars, which we cover with glass plates as soon as they are filled, we may examine the chemical change by which chlorine is formed. Manganese dioxide contains two atoms of oxygen, and this oxygen combines with the hydrogen of the hydrochloric acid, forming water. Two atoms of oxygen require four atoms of hydrogen, and for these four atoms we will need four molecules of hydrochloric acid, each of which contains one atom of chlorine and one of hydrogen. The atom of manganese combines with two atoms of chlorine, forming a body called manganese chloride, and as there were four chlorine atoms in the hydrochloric acid, two of these will pass off as gas. We may write the reaction,

$$MnO^2 \quad + \quad 4HCl \quad = \quad MnCl^2 \quad + 2H^2O + Cl^2$$

Manganese dioxide.   Hydrochloric acid.   Manganese chloride.   Chlorine.

72. Properties.—Chlorine is a greenish-yellow gas, having an unpleasant, suffocating odor. We must be careful not to breathe it in a too undiluted form, for it causes violent coughing, and irritates the lungs. It is 1.247 times as heavy as an equal bulk of air, or 35.5 times as heavy as an equal volume of hydrogen. Its atomic weight is also 35.5: there are many other elements whose atomic weights and densities (when in the form of gas) compared to hydrogen are the same, and we must suppose that the molecules of such elements are like those of hydrogen in that each contains two atoms (§ 46). Chlorine dissolves in water: at ordinary temperatures, one litre of water will dissolve about two and a half litres of the gas. It may be liquefied at 15° by a pressure of four atmospheres, that is, four times as great as the ordinary pressure of the air.

Chlorine possesses very great affinity for the other elements, and the compounds which it forms with them are called chlorides.

Over one of our jars of the gas we place a piece of coarse wire gauze, through which we sprinkle some finely-powdered antimony. Each little particle burns, and we have a shower of fire, while a heavy cloud of white smoke settles in the jar: this smoke is antimony chloride. Into another jar we throw some pieces of Dutch leaf, a very thin brass used for cheap gilding: this also burns, and the copper and zinc of which the Dutch metal was composed are converted into copper chloride and zinc chloride. We may burn some thin copper in the same manner. We put a small piece of phosphorus in a deflagrating-spoon, and lower this into another jar: it burns with a pale flame into phosphorus chloride.

73. Of all the elements, hydrogen is that for which chlorine possesses the most remarkable affinities. In a room lighted only by a candle, we have mixed over salt water equal volumes of chlorine and hydrogen, and, after drying this mixture by passing it through a tube containing pumice-stone and sulphuric acid, we have filled with it some little bulbs, blown on thin glass tubes, and then sealed the ends of the tubes with little plugs of paraffin. It is easy to fill the bulb; we connect one end of it by a rubber tube on which is a pinch (A), to the tube of the bell-jar in which the mixture is made; then on pressing the jar into the salt water and loosing the pinch, the gas is forced through the bulb (Fig. 38). We keep these bulbs carefully covered from the

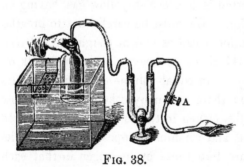

FIG. 38.

light. We now uncover one, put it behind a sheet of glass, and then, standing at a little distance, burn a piece of magnesium wire. Instantly there is an explosion; the hydrogen and chlorine have combined. The combination is brought about in the same manner by direct sunlight, and more gradually by diffuse daylight.

Chlorine does not support ordinary combustion, for it does not combine directly with carbon, and ordinary combustibles contain hydrogen and carbon; but their hydrogen may burn in chlorine. We put a lighted taper into a jar of chlorine, and the flame becomes red and smoky: the chlorine combines with the hydrogen of the wax, but the carbon separates in the form of smoke.

Chlorine even decomposes many compounds containing hydrogen, taking away that element to form hydrochloric acid. The solution of chlorine in water is decomposed by sunlight, the oxygen being set free.

$$2Cl^2 + 2H^2O = 4HCl + O^2$$

Into a jar with straight sides, filled with chlorine, we rapidly introduce a paper saturated with turpentine, and quickly replace the cover of the jar. There is a flash of red light, and a cloud of smoke. Turpentine is a compound of carbon and hydrogen only: the chlorine combines with the hydrogen, and the carbon forms the smoke. We find after the experiment that the paper is not burned: we use a plain straight jar, because it is easily cleaned by rubbing with a little turpentine.

We pour some blue litmus-water into a jar of chlorine; the blue liquid becomes colorless. In another jar we suspend a piece of moist colored calico, and it quickly fades. Chlorine possesses bleaching properties, and these properties are due to the decomposition of the dye-stuffs, nearly all of which contain hydrogen that the chlorine may remove. For the same reason chlorine is a valuable disinfectant, for most unpleasant and unwholesome odorous matters are compounds of hydrogen.

74. Chlorides.—The binary compounds of chlorine are called chlorides, and the same prefixes which are used for the names of the oxides are employed also to indicate the number of chlorine atoms in a molecule of the compound; thus, phosphorus trichloride contains PCl^3. In general, these prefixes are used to indicate the number of atoms of the second named element with which one or more atoms of that first named are combined. When there are only two chlorides which are important, the terminations *ous* and *ic* designate which contains the greatest and the least proportion of chlorine (see § 64). Mercurous chloride is Hg^2Cl^2; mercuric chloride is $HgCl^2$: this nomenclature also is of general application.

Among the chlorides of the non-metallic elements, hydrochloric acid is the most important.

The metallic chlorides are all soluble in water, with the exception of silver chloride, $AgCl$, mercurous chloride, Hg^2Cl^2, and cuprous chloride, Cu^2Cl^2. Lead chloride is only slightly soluble. Some chlorides are decomposed by water, and in such a case part or all of the chlorine combines with hydrogen, forming hydrochloric acid. Thus, phosphorous chloride, PCl^3, yields phosphorous acid and hydrochloric acid.

$$2PCl^3 \qquad + 3H^2O = \qquad 2H^3PO^3 \quad + \qquad 6HCl$$
Phosphorous chloride.   Phosphorous acid.   Hydrochloric acid.

75. We pour a few drops of a solution of silver nitrate in pure water into some water in which common salt, which is sodium chloride, has been dissolved. At once a white precipitate forms, for, while sodium nitrate now exists in solution, silver chloride is formed, and this is insoluble.

$$AgNO^3 \quad + \qquad NaCl \qquad = \qquad NaNO^3 \quad + \qquad AgCl$$
Silver nitrate.   Sodium chloride.   Sodium nitrate.   Silver chloride.

The precipitate darkens on exposure to light, and this reaction enables us to determine whether a body contains or does not contain a chloride. All solutions of chlorides give the white precipitate, which, we may add, dissolves if we pour off most of the liquid and then shake the white powder with strong ammonia-water.

LESSON IX.

HYDROCHLORIC ACID.—ACIDS.—SALTS.

76. Hydrochloric Acid, HCl.—We have seen that this compound is formed by the direct union of chlorine and hydrogen, and by the action of water on certain chlorides. Many chlorides are decomposed by water at high temperatures, and in this manner some mineral chlorides existing in the rocks cause hydrochloric acid to be formed in certain volcanic regions, where it mixes with the other gases that are emitted.

77. Preparation.—Hydrochloric acid is made by the action of sulphuric acid on common salt, the sodium of the salt changing places with the hydrogen of the sulphuric acid.

We put some pieces of rock-salt in a flask like that in which we made chlorine, and, if we wish a solution of the gas, we connect our delivery-tubes with a series of bottles containing water, through which the gas will be forced to bubble (Fig. 39). If we

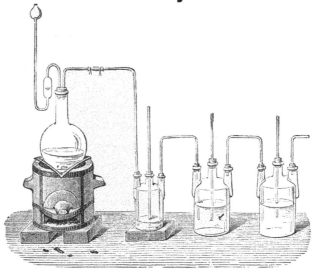

FIG. 39.

wish the dry gas, we dry it as we did the chlorine. When all is ready, we pour through the safety-tube sulphuric acid which we have previously diluted with an equal volume of water, and immediately the gas begins to come off. When the reaction becomes tranquil, we must heat the mixture.

One molecule of sulphuric acid contains two atoms of hydrogen, and may be made to yield one or two molecules of hydrochloric acid, by reacting with one or with two molecules of salt.

$$NaCl \quad + \quad H^2SO^4 \quad = \quad HCl \quad + \quad NaHSO^4$$
Sodium chloride.　Sulphuric acid.　Hydrochloric acid.　Sodium acid sulphate.

$$2NaCl + H^2SO^4 = 2HCl + \quad Na^2SO^4$$
Sodium sulphate.

This reaction is operated on an enormous scale in Europe, where the sodium sulphate is afterwards heated with chalk and converted into sodium carbonate.

78. Properties.—Hydrochloric acid is a colorless, pungent, and suffocating gas. Its density compared to hydrogen is 18.33, sufficiently near that which would be indicated by half its molecular weight, which is 36.5 (see § 48). It is very soluble in water, and if under the surface of water we remove the cork from a bottle filled with the gas, the water at once rises and fills the bottle. At 0° one litre of water will dissolve 500 litres of hydrochloric acid. The strongest hydrochloric acid of commerce, commonly called muriatic acid, contains about 34 per cent. of the gas. Like the gas, it produces fumes in the air by condensing the moisture in the atmosphere.

Hydrochloric acid is a strong acid. A drop or two of the solution will redden a large quantity of blue litmus. We slowly pour some hydrochloric acid into a strong solution of sodium hydrate : a white powder soon separates, and we can satisfy ourselves by tasting it that this is common salt. Water also is formed.

$$NaOH + HCl = H_2O + NaCl$$
Sodium hydrate.　　　　　　　　　　　　Sodium chloride.

With oxides of the metals, hydrochloric acid acts in the same manner, water and a chloride being formed.

$$HgO + 2HCl = HgCl_2 + H_2O$$
Mercuric oxide.　　　　　Mercuric chloride.

We have seen that zinc liberates the hydrogen from hydrochloric acid : many other metals act likewise.

ACIDS AND SALTS.

79. An acid is a compound containing hydrogen which is capable of being replaced by a metal, forming a body which is called a salt. Although salts may be formed in various manners, we have an exact definition : a salt represents an acid whose hydrogen has been partly or wholly replaced by metal. Hydrochloric acid is an example of a binary acid, but the few binary acids which we shall study have not all as energetic properties as hydrochloric acid. The salts formed by hydrochloric acid are of course the chlorides.

80. **Hypochlorous Oxide and Acid.**—When chlorine is passed over cooled mercuric oxide, mercuric chloride is formed, and the oxygen which separates from the mercury combines• with chlorine, forming a gas which may be condensed to a yellow liquid by passing it into a bottle surrounded by a freezing mixture of ice and salt.

$$HgO \quad + \quad 2Cl^2 = \quad HgCl^2 \quad + \quad Cl^2O$$
Mercuric oxide.   Mercuric chloride.   Hypochlorous oxide.

This is hypochlorous oxide; it is a dangerous body, and often explodes without warning.

It reacts with water in a manner which we must study. A molecule of the oxide and a molecule of water interchange a chlorine. atom for a hydrogen atom, and a compound called hypochlorous acid is formed.

$$ClOCl \quad + HOH = \quad HOCl \quad + \quad ClOH$$
Hypochlorous oxide.   Water.   Hypochlorous acid.   Hypochlorous acid.

This is an oxygen acid, and we may consider that it is com· posed of an atom of chlorine combined with the residue of a molecule of water from which one atom of hydrogen has been removed. This residue would be OH, and, because the atom of oxygen has not enough hydrogen to satisfy the affinities, it is not a molecule; it cannot exist except as part of a molecule; that is, combined with some other atom. It is called, for convenience' sake,·*hydroxyl*, and all oxygen acids contain this group of two atoms, hydroxyl. Indeed, all of the compounds we call hydrates contain the group hydroxyl: thus, potasssium hydrate is KOH.

81. We have had occasion to notice the names hydrochloric acid, hypochlorous acid, sulphuric acid. We have seen that hydrochloric acid produces binary salts: the names of binary compounds, with the exception of acids, end in *ide*, and we can now even understand that a sulphide is a compound containing sulphur and one other element. But hypochlorous acid and sulphuric acid are not binary compounds; they may be formed respectively by the action of hypochlorous and sulphuric oxides on water. The first of these actions we have studied: the second we may write

$$SO^3 + H^2O = H^2SO^4.$$
6*

When the hydrogen of either of these oxygen acids is replaced by metal, how shall we name the resulting salts?   Chemists have agreed that the termination *ic* shall be changed to *ate*, and *ous* shall be changed to *ite*.   This is a simple nomenclature.   The salts of sulphuric acid must be sulphates; those of nitric acid, nitrates; those of permanganic acid, permanganates; those of hypochlorous acid, hypochlorites; those of sulphurous acid, sulphites.   We see also that the chlorates must be the salts of chloric acid; the arsenites, those of arsenious acid.

82. Hypochlorites.—Solutions of the hypochlorites of potassium and sodium are useful as disinfecting and bleaching liquids. They are made by passing chlorine gas into a rather dilute solution of potassium hydrate or sodium hydrate; at the same time water is formed, and a chloride, which remains in solution.

$$2NaOH \quad + Cl^2 = \quad NaOCl \quad + \quad NaCl \quad + H^2O$$

Sodium hydrate.   Sodium hypochlorite.   Sodium chloride.

Such a liquid quickly removes the stains of wine and fruits from linen, and also deodorizes offensive matters.

Bleaching powder, or chlorinated lime, is made by passing chlorine gas over slaked lime.   Its solutions contain calcium hypochlorite, $Ca(ClO)^2$, and may be substituted for the liquids which we have just mentioned.   The bleaching and disinfecting by these substances are due to their decomposition, which we may suppose first sets free hypochlorous acid, and this attacks the coloring matter or offensive substance, removing hydrogen; the chlorine atom will take one atom of hydrogen, forming hydrochloric acid, and the group OH takes another atom, forming water.   We may understand this by examining the reaction between hydrochloric and hypochlorous acids, which yields chlorine and water.

$$HClO + HCl = H^2O + Cl^2$$

83. Chlorates.—We pass a current of chlorine gas into a strong solution of potassium hydrate, and a white solid matter soon appears in the liquid; when this no longer increases in bulk, we stop the chlorine, heat the liquid until it boils, and if all of the solid dissolves, we evaporate it until a considerable quantity of this matter again separates.   We now allow it to settle a moment, and pour off the clear liquid : as this cools, shining little crystals separate in rhomboidal plates. These are potassium chlorate, and we have been obliged to separate them from potassium chloride, which, together with water, is also formed during the experiment.

$$6KOH + 3Cl^2 = 5KCl + KClO^3 + 3H^2O$$

Potassium hydrate.　　　　　Potassium chloride.　　Potassium chlorate.

Potassium chlorate is the most important salt of chloric acid, $HClO^3$, which we might prepare from the salt by a troublesome process. Potassium chlorate is not very soluble in cold water, but is very soluble in boiling water.• Its solution is excellent as a gargle for sore throat, but must not be swallowed, for it is poisonous. We have seen that potassium chlorate is decomposed by heat, yielding oxygen : it readily gives up its oxygen, for the chlorine has a much stronger affinity for the potassium than for the oxygen, which appears to hold the chlorine and potassium atoms together.

84. Chloric acid, which would be set free by the action of stronger acids on potassium chlorate, is at once decomposed under such circumstances if oxidizable substances be present. On a mixture of equal parts of potassium chlorate and sugar, powdered separately, we let fall a drop of strong sulphuric acid. The mixture at once takes fire and burns vividly, the potassium chlorate furnishing the oxygen for the combustion of the sugar.

Into a tall jar, filled with water, we throw some crystals of potassium chlorate, and on them a small piece of phosphorus ; then, by means of a funnel-tube which passes to the bottom of the jar, we pour some strong sulphuric acid on the chlorate. The chloric acid set free is decomposed by the phosphorus and causes its combustion under the water (Fig. 40).

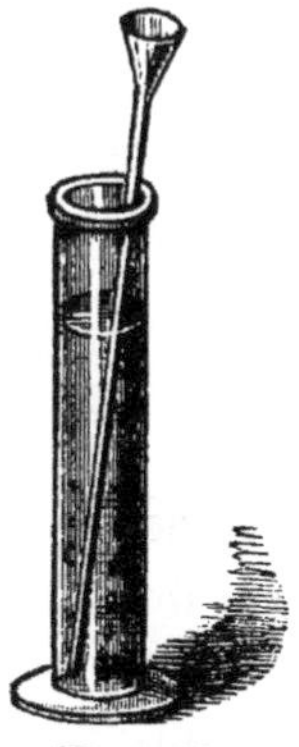

FIG. 40.

We put into a mortar a piece of sulphur about as large as a match-head, and a crystal of potassium chlorate of the same size ; then we rub them briskly together, being careful to keep the mortar far enough from the face, and soon there is a loud report; the sulphur has been oxidized and the potassium chlorate decomposed. If we used larger quantities of these substances in our experiment, we might break the mortar, and possibly injure our person.

LESSON X.

BROMINE.—IODINE.—FLUORINE.

85. **Bromine,** Br $=$ 80.—In a long tube closed at one end, we dissolve in a little water a few crystals of a white substance, called potassium bromide, and then pour in some chlorine-water, which we have prepared by passing chlorine through water contained in bottles such as were used in the preparation of the solution of hydrochloric acid: we now add a considerable proportion of ether, and shake the tube after closing it with the finger. The liquid becomes brown, and after standing a few minutes, the ether, which is not very soluble in water, comes to the surface, and its color is red, while the water has become colorless. The potassium bromide, a compound of potassium and bromine, has been decomposed by the chlorine, and potassium chloride formed in the solution, while the bromine set free has been dissolved by the ether, in which it is much more soluble than in water. We may write the reaction,

$$2KBr \quad + Cl^2 \quad = \quad 2KCl \quad + \quad Br^2$$
Potassium bromide.　　Potassium chloride.　　Bromine.

86. The compounds of bromine with potassium, sodium, and magnesium, which compounds are called bromides of those metals, are found in the waters of many salt springs, and exist in small quantity in the water of the ocean. As they are much more soluble in water than common salt, they remain dissolved when most of the salt has been separated by evaporating the liquid, and from their concentrated solution so obtained the bromine is separated by heating the liquid with sulphuric acid and manganese

dioxide. Supposing all of the bromine to exist as potassium bromide, manganese sulphate, potassium sulphate, and water are formed at the same time, while the bromine distils, and is condensed in suitable apparatus.

$$2KBr + MnO^2 + 2H^2SO^4 = K^2SO^4 + MnSO^4 + H^2O + Br^2$$

| Potassium bromide. | Manganese dioxide. | Sulphuric acid. | Potassium sulphate. | Manganese sulphate. |

87. Bromine is a dark-red liquid, having an exceedingly irritating odor. Its density is 2.99. It freezes at —24°, and boils at 63°; it is very volatile at ordinary temperatures. It dissolves in about thirty times its weight of water at 15°, and is quite soluble in ether, chloroform, and carbon disulphide, liquids which dissolve many substances that are not soluble in water.

Bromine closely resembles chlorine in its chemical reactions, but its affinities are not as powerful. Its solution in water will bleach litmus, and other coloring matters, but more feebly than chlorine. We pour a little bromine into a deep test-tube and drop in a small piece of warm thin sheet copper, which is instantly converted into copper bromide with the production of heat and light.

Bromine combines with hydrogen, forming hydrobromic acid, HBr, a gas which dissolves in water, and undergoes reactions similar to those of hydrochloric acid.

Bromine is exceedingly corrosive to animal tissues, and is sometimes employed as a caustic in surgery. It also disinfects like chlorine.

The atomic weight of bromine is 80, and the density of its vapor compared to hydrogen is also 80, showing that a molecule of bromine contains two atoms.

88. **Iodine, I = 127.**—In a tube like that which we used in the experiment with potassium bromide, we dissolve a little potassium iodide in water, add chlorine-water as before, and then, instead of ether, we pour in some carbon disulphide. After shaking the tube, and allowing it to stand, the carbon disulphide, being heavier than the water, is found at the bottom with a beautiful purple color. Were we to pour off the watery liquid and allow

this carbon disulphide to evaporate in a shallow dish, it would leave a brownish-gray matter, which is iodine, and which the chlorine has driven out of the potassium iodide, just as it separated the bromine from the potassium bromide.

89. Like bromine, iodine is found combined with potasssium, sodium, and magnesium in the waters of some springs, and in sea-water. It also exists in small quantity in the sodium nitrate found in large deposits in Chili, and, being very soluble in water, remains in the *mother-liquor*, as it is called, from which this sodium nitrate has been crystallized for its purification. It is obtained from these liquids, and from the ashes of sea-weeds; the sea-weeds are burned, and the iodides which are dissolved out of the ashes by water, are treated just as the bromides are treated for the preparation of bromine. Iodine may also be separated by adding nitric acid to the solution of an iodide, and we may make the experiment by pouring a little nitric acid on some potassium iodide solution in a test-tube. Potassium nitrate is formed, and iodine deposits as a dark powder: the red vapors that are given off are a compound of nitrogen and oxygen, which we will study in good time.

90. Iodine is purified by sublimation; that is, heating it, and condensing the vapor. When pure, it is in crystalline, bluish-gray plates, much like scales of plumbago. Its density is 4.95. It melts at 107°, and boils at 175°. We carefully heat a few small scales of iodine in a large glass flask, which soon becomes filled with a magnificent purple vapor. This vapor is so heavy that we may pour it out on a piece of cold glass, where it condenses in minute crystals. The density of this vapor compared to hydrogen is 127, and, the atomic weight being 127, we see that the iodine molecule contains two atoms.

Iodine is very slightly soluble in water: one part of iodine requires 7000 parts of water to dissolve it, and yet the solution has a distinct brown color. It dissolves readily in alcohol, ether, chloroform, and carbon disulphide, and the color of the solution depends on the solvent; that in alcohol is brown, but that in chloroform is violet.

91. We have made some thin starch paste by boiling starch with water, and we pour some of this into two test-tubes : to the first we add a few drops of a solution of iodine in water, and the liquid becomes dark blue; to the other we add a drop or two of a solution of potassium iodide, and no color is produced. Starch is dyed a blue color by iodine, but the iodine must be free; on adding a few drops of chlorine-water to the second tube the potassium is removed from the iodine, and the blue color at once appears. This is the test for iodine.

Iodine combines with hydrogen to form hydriodic acid, HI, a gas whose properties are much like those of hydrochloric acid. It is made by heating water with iodine and amorphous phosphorus.

92. **Analogies of Cl, Br, and I.**—On comparing the three elements which we have just considered, we find that while one is a gas, another liquid, and the third solid, still the corresponding compounds formed by each are much alike in chemical nature; that is, the composition and reactions of the molecule. The compounds with hydrogen each contain one atom of hydrogen combined with one of the other element: if the power to combine with one atom of hydrogen be taken as the measure of the combining power of any atom, the atoms of chlorine, bromine, and iodine must have equal powers. Since an atom of each of these elements combines with only one atom of hydrogen, they are said to be *monatomic* elements in their compounds with hydrogen. But their affinities, or energies of combination, for hydrogen are not alike: chlorine will take the hydrogen away from hydrochloric acid, and bromine will take it away from hydriodic acid. This is also the order of their affinity for the metals, but in the *number of atoms* of either chlorine, bromine, or iodine which will combine with one atom of another element, the three are exactly alike.

In this last respect the next element resembles the three which we have just studied.

93. **Fluorine,** Fl.—We have evenly coated one side of a glass plate with wax, and in the wax we trace a design with a sharp point, taking care that our lines go quite through to the glass. In a shallow dish made of sheet lead, we mix, by the aid of a wooden stick, some powdered fluor-spar, which is a mineral, with strong sulphuric acid; over this we place our glass containing the design, with the waxed side down, and we gently warm the dish (Fig. 41). In a few minutes we remove the glass, and, after

gently warming it, rub off the wax: we find that the design is
permanently etched into the glass.   The fluor-spar is a compound
of the elements fluorine and calcium, and the sulphuric acid has

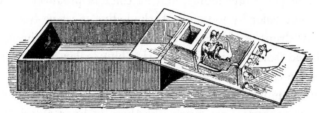

Fig. 41.

decomposed it, forming a vapor called *hydrofluoric acid*, a compound of hydrogen and fluorine.

$$CaFl^2 \quad + \quad H^2SO^4 \quad = \quad CaSO^4 \quad + \quad 2HFl$$
Calcium fluoride.   Sulphuric acid.   Calcium sulphate.   Hydrofluoric acid.

Hydrofluoric acid may be condensed to a liquid, and it may be
dissolved in water, but neither the liquid nor the solution can be
kept in glass bottles, because fluorine has an extraordinary affinity
for the element silicon which forms part of the glass, and it would
combine with that element, destroying both bottle and acid.   It
is to this affinity that we owe the etching of our glass plate.
Bottles of india-rubber or of lead are used to contain hydrofluoric
acid, for it does not attack those substances.   The graduations on
delicate chemical apparatus, such as the eudiometers we have seen,
are etched into the glass by this acid.   Hydrofluoric acid is very
corrosive, and we must be careful in its use.

The powerful affinities of fluorine have thus far prevented chemists from
separating the element and studying its properties; but, since we know that the
molecular weight of hydrofluoric acid is 20, and have reason to believe that
the molecule contains only one atom of hydrogen, we conclude that the atomic
weight of fluorine is 19.

Besides fluor-spar, there is another important compound of fluorine found
in nature; it is the mineral *cryolite*, which is a double compound of sodium
fluoride and aluminium fluoride.

LESSON XI.

SULPHUR.—HYDROGEN SULPHIDE.

94. Sulphur, S $= 32$.—We are all familiar with sulphur, or brimstone. In some localities it is found pure or very impure and mixed with the soil: especially is this the case in volcanic countries. Besides this free or *native* sulphur, as it is called, sul-

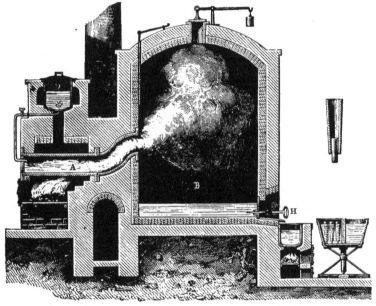

Fig. 42

phur is found combined with many metals, and the compounds are called sulphides.

Crude sulphur comes in large quantities from Sicily, where it is obtained by distilling it from the earthy matters with which it is mixed. It is refined by again distilling it in an apparatus consisting of an iron boiler (A, Fig. 42), above which is a reservoir (C) where the sulphur is first melted by the waste heat, and from which it runs into the boiler. The sulphur vapor enters a large

D 7

chamber (B), and after condensing runs down on the floor, which is inclined so that the melted sulphur may be drawn off at a tap (H). While the walls of this chamber are yet cold, the sulphur condenses in a fine yellow powder, which is sold as *flowers of sulphur;* but when the chamber becomes heated, the condensed sulphur melts, and after being drawn from the opening is cast in cylindrical moulds, in which it solidifies and becomes *roll sulphur.*

Large quantities of sulphur are also obtained by distilling *iron pyrites,* a compound which contains iron and sulphur, and which gives up part of its sulphur when it is heated.

95. PROPERTIES.—Sulphur is a brittle, lemon-yellow solid, having neither taste nor odor. It is a bad conductor of electricity and heat : a piece of roll sulphur held firmly in the hand produces a curious crackling noise, because the outside becomes warm, and its expansion causes it to crack before the heat can be conducted to the interior. Sulphur is not soluble in water, is very slightly soluble in alcohol and ether, but dissolves readily in carbon disulphide. When heated, it melts at 111°, and becomes a mobile, brown, and transparent liquid.

We melt some sulphur in an earthen crucible, and, as soon as it has all melted, we allow it to cool until a crust forms over the surface. We now make a hole in the crust, and pour out the sulphur which has not solidified. On breaking off the crust, we find the interior of the crucible lined with beautiful, transparent crystals, which on close examination we determine to be oblique rhombic prisms. In a glass flask we melt some more sulphur, but after it has melted we keep on heating it: we see that the color becomes darker, and the liquid thicker. When its temperature reaches 220°, we can turn the flask upside down and the sulphur scarcely runs on the sides. At about 260° it again becomes liquid, and as soon as we see it becoming more fluid, we pour it into a vessel of cold water, moving the flask so that all does not fall in the same place. On taking the sulphur from the water, we find that its properties are much changed: it is transparent and very elastic; we pull it out in long threads. This curious form, which is called *soft sulphur,* is due to a molecular condition of the element; we must believe that its molecules contain more energy than those of ordinary sulphur, for if we gradually heat it, it at once becomes opaque and brittle, and at the same time much hotter than we have heated it. It changes spontaneously in this manner after we have kept it a few hours. Soft sulphur is amorphous; that is, has no crystalline form.

Besides these two forms of sulphur, prismatic crystals and soft sulphur, there is another. When it is found crystallized in nature, the crystals are right rhombic octahedra. After a time the crystals in our crucible become opaque and brittle: when we examine them, we find that each little prism has separated into several octahedra, the faces of which hold together until we break them apart.

When a solution of sulphur in carbon disulphide is evaporated, the sulphur is deposited in the octahedral form, and at ordinary temperatures the other forms gradually change to this. Because sulphur has more than one distinct physical form, it is said to be *dimorphous*. The density of prismatic sulphur is 1.98; that of octahedral sulphur is 2.05.

96. Sulphur takes fire in the air at a temperature below redness: its combustion is its union with oxygen, forming sulphur dioxide, SO^2, called also sulphurous oxide and sulphurous acid gas. By the aid of heat, sulphur unites directly with many other elements: we have seen in one of our experiments (§ 4) that copper burns brilliantly in sulphur vapor, and in the same manner we might burn some iron wire, forming iron sulphide.

Sulphur is used in the manufacture of matches, gunpowder, sulphuric acid, and many other operations.

97. **Sulphides.**—We put a little antimony sulphide into a test-tube, and boil it with some hydrochloric acid. A gas having the unpleasant odor of rotten eggs is given off, and antimony chloride remains in the tube. This gas, which we shall now study, is called hydrogen sulphide, or sulphuretted hydrogen; nearly all the sulphides form this gas when boiled with hydrochloric acid, and the reaction gives us a test for the sulphides.

98. **Hydrogen Sulphide,** H^2S.—Into a bottle like that which

served for the preparation of hydrogen, we put some ferrous sulphide, which we have made by heating to redness in an earthen crucible a mixture of iron filings with about its own weight of sulphur. We then pour through the funnel-tube some dilute sulphuric acid, and at once or in a few minutes an effervescence shows us that gas is being given off, and we soon detect this gas by its odor. It is a compound of hydrogen and sulphur, and is formed by the reaction

$$FeS \quad + \quad H_2SO_4 \quad = \quad H_2S \quad + \quad FeSO_4$$

Ferrous sulphide.   Sulphuric acid.   Hydrogen sulphide.   Ferrous sulphate.

The ferrous sulphate formed remains dissolved in the water.

As we often desire this gas in the laboratory, we sometimes employ a self-regulating apparatus consisting of two bottles which

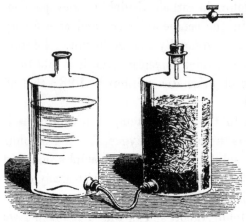

have openings near the bottom, and these openings are connected by a stout rubber tube (Fig. 43). In one we put a layer of clean pebbles that rise above the lower opening, and on this the ferrous sulphide; to the neck of this bottle we adapt a glass stop-cock by the aid of a good cork. In the other bottle, which

FIG. 43.

we must not cork, we pour our dilute sulphuric acid. When we open the stop-cock, the acid runs in on the ferrous sulphide; the gas is then formed, and we may keep it in the bottle or use it as we desire: when we close the stop-cock, the gas forming in the bottle forces the acid into the other bottle, and as soon as the surface of the acid is below the top of the layer of pebbles, the ferrous sulphide is no longer acted on. We may use this apparatus for the preparation of hydrogen and carbon dioxide (§ 234), of course cleaning it out before changing the materials.

99. PROPERTIES.—Hydrogen sulphide is a colorless gas, having

a stinking and penetrating odor. Its density being 17 times that
of hydrogen or $17 \times .0693$ that of the air, its molecular weight
must be 34 (§ 48). By strong pressure it is converted into a colorless liquid. At ordinary temperatures water dissolves about three
times its volume of hydrogen sulphide, and the solution is sometimes used in the laboratory, but it does not keep long, for the air
oxidizes the hydrogen, forming water, while sulphur is deposited.

We can easily determine the composition of this gas. Into a long test-tube
of hard glass we thrust a roll of tin foil, and, after tightly corking the tube,
we heat it until the tin acquires a yellow color. After the tube has cooled, we
uncork it under the surface of mercury, and we find that the volume of gas
has not changed. This gas is hydrogen, and two volumes (one molecule) of
hydrogen sulphide therefore contain two volumes (two atoms) of hydrogen.
If from the density (half the molecular weight) of hydrogen sulphide we subtract that of hydrogen, we obtain $(17 - 1 = 16)$ a number equal to the weight
of half the sulphur in a molecule of the compound. The sulphur in a molecule of hydrogen sulphide therefore weighs 32 if the hydrogen weighs 2.

Hydrogen sulphide is a combustible gas, as we can easily understand, since its molecule contains only hydrogen and sulphur, both
of which are able to unite with the oxygen of the air, the first to
form water, and the second to form sulphur dioxide, the same gas
which is formed when sulphur burns in the air. When we light
the gas at the end of the delivery-tube, it burns with a blue flame.

100. Certain reactions of this gas make it exceedingly valuable
in the laboratory. We pass the delivery-tube from our apparatus
into a solution of copper sulphate in water: a brownish-black precipitate is formed as soon as the gas comes in contact with the
liquid. This is copper sulphide, and sulphuric acid remains in the
solution.

$$CuSO^4 \quad + \quad H^2S \quad = \quad CuS \quad + \quad H^2SO^4$$
Copper sulphate.   Copper sulphide.   Sulphuric acid.

We pass the gas into a solution of antimony chloride, and an
orange-colored precipitate of antimony sulphide forms, while hydrochloric acid is in the liquid.

$$2SbCl^3 \quad + \quad 3H^2S \quad = \quad Sb^2S^3 \quad + \quad 6HCl$$
Antimony chloride.   Antimony sulphide.

In a solution of zinc acetate, we would have thrown down a
white precipitate of zinc sulphide. Naturally, in these reactions

we must know by analysis the composition of the molecules which react together, and that of the bodies which are formed, before we can write the equations. The solutions of many other metallic compounds are decomposed in this manner by hydrogen sulphide, and the color and other properties of the metallic sulphide formed show us what metal exists in the solution to which we apply the test.

Hydrogen sulphide is at once decomposed by chlorine, hydrochloric acid being set free.

$$H^2S \quad + \quad Cl^2 \quad = \quad 2HCl \quad + \quad S$$

Hydrogen sulphide is a poisonous gas, and must not be inhaled for any length of time, even when very much diluted with air.

101. Sulphydrates.—We have seen that hydrates are formed by the action of water on the oxides (§ 62), and that these hydrates contain the group of atoms OH, which we call hydroxyl. On examining the composition of the molecule of hydrogen sulphide, we see that it is exactly like that of water, but contains a sulphur atom instead of an oxygen atom.

HOH   HSH
Water.   Hydrogen sulphide.

There are also compounds exactly like the hydrates, but containing sulphur instead of oxygen, and they are called sulphydrates. We pass hydrogen sulphide into a solution of potassium hydrate; it is absorbed, and a chemical reaction which takes place causes the liquid to become warm.

KOH   +   HSH   =   KSH   + HOH
Potassium hydrate.   Hydrogen sulphide.   Potassium sulphydrate.   Water.

We cannot fail to notice the similarity between the structure of these molecules, and this similarity leads us to the conclusion that as far as combining with atoms of hydrogen and potassium is concerned, there must be a resemblance between sulphur atoms and oxygen atoms. We will in time notice that this resemblance does not stop here, but is borne out in the structure of many other molecules containing sulphur and oxygen atoms. Since one atom of sulphur or one of oxygen is capable of combining with two atoms of hydrogen or one of hydrogen and one of potassium, and since we take the hydrogen atom as the unit of the combining power, we say that the hydrogen and potassium atoms are mono-atomic, and that the oxygen and sulphur atoms in these compounds are diatomic.

LESSON XII.

SULPHUR DIOXIDE.—SULPHUR TRIOXIDE.

102. Sulphur Dioxide, SO^2.—This compound is formed when sulphur burns in the air or in oxygen: we could not obtain it pure by burning sulphur in air, for it would then be mixed with the other constituents of the air. We usually prepare the gas by boiling sulphuric acid with copper clippings: the products of the operation are cupric sulphate, water, and sulphur dioxide: knowing this, we may write our equation,

$$\overset{.}{C}u \quad + \quad 2H^2SO^4 \quad = \quad CuSO^4 \quad + \quad 2H^2O \quad + \quad SO^2$$
Copper.   Sulphuric acid.   Cupric sulphate.   Sulphur dioxide.

We conduct the experiment in an apparatus like that in which we prepared chlorine, and if we desire to collect the gas we do so by downward dry displacement.

103. PROPERTIES.—Sulphur dioxide is a colorless, suffocating gas. Its density compared to hydrogen is 32, agreeing with that which our theory should indicate (§ 48), and it is therefore a little more than twice as heavy as an equal volume of air. By pressure, or by a temperature of $-10°$, it is converted into a colorless liquid, and this liquid may be easily prepared by passing the gas into a bottle surrounded by a mixture of ice and salt. The evaporation of the liquid produces great cold: a temperature as low as $-73°$ has been obtained by aiding this evaporation by pumps, and the phenomenon has been applied in the construction of certain machines for making ice. We can easily freeze some water in a test-tube by wrapping the lower end of the tube in some cotton wool, and pouring on this some liquid sulphurous oxide, but we must make the experiment in a current of air to carry off the suffocating gas.

At ordinary temperatures water dissolves about forty times its volume of sulphurous oxide, and the solution is frequently employed in the laboratory.

104. Sulphurous oxide is naturally not combustible, for the sulphur which it contains has had an opportunity to combine with all of the oxygen with which it would unite. It extinguishes burning bodies.

While, however, one atom of sulphur will not combine directly with more than two atoms of oxygen, sulphurous oxide can be still further oxidized by certain reactions. If it be mixed with oxygen, and the mixture passed through a red-hot tube containing platinum sponge, the two gases combine, forming sulphur trioxide, SO^3; the vapor of this substance may be condensed by passing it into an ice-cold receiver. By the action of nitric acid, sulphur dioxide is converted into sulphuric acid, and the reaction is applied in the manufacture of the latter acid.

In a tall jar we dissolve some potassium permanganate in water; this body contains a large proportion of oxygen, with which it parts easily to oxidizable matters. We pass some sulphur dioxide through the purple solution, which is rapidly decolorized; the sulphur dioxide becomes sulphuric acid in this reaction, and the potassium permanganate is said to be reduced. We use the term *reduction* to mean taking away oxygen, and any body which is capable of removing oxygen from other substances is called a reducing agent.

Sulphur dioxide is used for bleaching wool, straw, and other matters which would be injured by chlorine. The substances are bleached by being put in a room in which sulphur is burned. We may in this manner bleach a flower in a bell-jar under which some sulphur is burning.

105. **Sulphites.**—When sulphur dioxide dissolves in water, the two substances really combine, and, by a reaction analogous to that which formed hypochlorous acid, sulphurous acid is formed. In this case, however, the two atoms of hydrogen exist in one molecule of the resulting acid.

$$SO^2 \quad + \quad H^2O \quad = \quad H^2SO^3$$
Sulphur dioxide.   Sulphurous acid.

Sulphurous acid is not a stable compound; it is decomposed when we try to separate it from its solution, and yields again sulphur dioxide and water. However, both of the hydrogen atoms are replaceable by metal, forming salts which are called sulphites; by passing sulphur dioxide into a solution of sodium hydrate, sodium sulphite and water are formed.

$$2\text{NaOH} \quad + \quad \text{SO}^2 \quad = \quad \text{Na}^2\text{SO}^3 \quad + \quad \text{H}^2\text{O}$$

Sodium hydrate.   Sodium sulphite.

To a little of this sodium sulphite in a test-tube we add hydrochloric acid; we can at once detect the pungent odor of sulphur dioxide, and a solution of common salt remains in the tube.

$$\text{Na}^2\text{SO}^3 + 2\text{HCl} = \text{SO}^2 + \text{H}^2\text{O} \cdot + 2\text{NaCl}$$

This gives us a test by which we may recognize a sulphite.

106. When a sulphite is boiled with sulphur, the latter is dissolved, and a compound called a *thiosulphate*, or formerly named *hyposulphite*, results. With sodium sulphite, we would have sodium thiosulphate.

$$\text{Na}^2\text{SO}^3 \cdot \quad + \quad \text{S} \quad = \quad \text{Na}^2\text{S}^2\text{O}^3 \text{ or Na}^2\text{SO}^3\text{S}$$

Sodium sulphite.   Sodium thiosulphate.

It will be noticed that the thiosulphate has exactly the composition of a sulphate ($\S$ 113) in which an atom of oxygen is replaced by an atom of sulphur. When a thiosulphate is treated with an acid, sulphur dioxide is evolved, and sulphur separates. The rags used in the manufacture of paper are bleached by chlorine, but no chlorine must be left in the paper, or this would be injured. In presence of water, sulphur dioxide (we may then say sulphurous acid) and chlorine react to form sulphuric and hydrochloric acids, both of which may readily be neutralized.

$$\text{H}^2\text{SO}^3 \quad + \quad \text{Cl}^2 \quad + \quad \text{H}^2\text{O} \quad = \quad \text{H}^2\text{SO}^4 \quad + \quad 2\text{HCl}$$

Sulphurous acid.   Sulphuric acid.

Sodium thiosulphate therefore serves as an *antichlor* in the manufacture of paper.

107. **Sulphur Trioxide, SO³.**—We have seen that this compound is formed by the direct union of sulphur dioxide and oxygen in presence of heated platinum sponge ($\S$ 104): other porous substances cause the same combination. Sulphur trioxide is usually prepared by heating fuming sulphuric acid, which is sometimes called *Nordhausen acid*, because it was for a time manufactured only in the village of Nordhausen, in Saxony, by distilling partially-dried ferrous sulphate. It is a compound of sulphur trioxide and sulphuric acid; $\text{H}^2\text{SO}^4 + \text{SO}^3 = \text{H}^2\text{S}^2\text{O}^7$. When it is heated, it decomposes into its constituents; the sulphur trioxide, being the most volatile, is condensed in cold flasks, which are at once hermetically sealed.

Sulphur trioxide is a snowy-white solid, crystallizing in feather-like flakes. It combines so energetically with water that each particle makes a hissing noise like hot iron on touching the liquid. The result of this combination is sulphuric acid.

$$\text{SO}^3 + \text{H}^2\text{O} = \text{H}^2\text{SO}^4$$

LESSON XIII.

SULPHURIC ACID, H_2SO_4.

108. Into a jar of oxygen containing a little water, we lower a deflagrating-spoon containing burning sulphur. The jar soon becomes filled with sulphur dioxide, and when the flame of the sulphur is extinguished, we pour into the jar a little nitric acid. Red vapors at once become apparent, but disappear in a little while: in order to mix the gases well, we now shake the jar, keeping it closely covered, and then by means of a long glass tube we blow some air into it: red vapors are again produced. It is evident that some chemical change has occurred between the nitric acid and sulphur dioxide, and that another change takes place between the gases in the jar and the air which we have introduced. The first change is the production of sulphuric acid, and the conversion of the nitric acid into the red vapors of nitrogen peroxide. Since one molecule of nitric acid contains only one atom of hydrogen, while one molecule of sulphuric acid contains two such atoms, two molecules of nitric acid must react with one of sulphur dioxide.

$$SO_2 + 2HNO_3 = H_2SO_4 + 2NO_2$$
          Nitric acid.   Sulphuric acid.   Red vapors.

The red vapors react with the water in the jar, and again yield nitric acid and another gas, nitrogen dioxide, which is colorless.

$$3NO_2 + H_2O = 2HNO_3 + NO$$
    Red vapors.   Nitric acid.   Nitrogen dioxide.

The nitric acid so regenerated oxidizes more sulphur dioxide, and this reaction continues until the gases in the jar are a mixture of nitrogen dioxide and sulphur dioxide. But the nitrogen dioxide is not lost: it takes an atom of oxygen from the air blown into the jar, and again forms red vapors.

$$NO + O = NO_2$$

These red vapors in turn react with the water, again forming nitric acid, which oxidizes more sulphur dioxide, and this series of reactions continues until all of the sulphur dioxide is converted into sulphuric acid.

109. These reactions are those which actually take place in the manufacture of sulphuric acid, which is commonly called oil of vitriol, and of which enormous quantities are used at one stage or another in the manufacture of nearly all other chemical compounds. The sulphur dioxide is obtained by burning sulphur in furnaces (AA, Fig. 44), the heat of which boils water for the steam required in the operation. The sulphur dioxide formed passes through a series of leaden chambers, in one of which (D) it comes in contact with nitric acid, which trickles down over a sort of cascade (EE). The gases then pass through other leaden chambers into which steam is injected (HH) for the regeneration of nitric acid, and the sulphuric acid formed collects on the floor of the chambers, from which it is drawn off. An excess of air must be passed into the chambers in order to reoxidize the nitrogen dioxide, and as the nitrogen of the air which must be allowed to escape from the apparatus would carry off some of that oxide of nitrogen, all the waste gases are obliged to pass through a tower (R) filled with coke which is kept wet with strong sulphuric acid. This latter absorbs the nitrous gases, and as it runs from the tower is conducted into a vessel (i), from which it may be forced by steam pressure to the top of the first small chamber (C), through which the sulphur dioxide is caused to pass. Here the sulphur dioxide removes all of the nitrous gases and carries them again into the chambers, so that only nitrogen from the air used escapes at the chimney of the coke column. The chambers are of various sizes, sometimes five metres wide and high, and ten, twenty, or even more metres in length.

The acid drawn from the leaden chambers is called *chamber acid:* it is strong enough for many purposes, its density being 1.5. The strong acid, density 1.842, is made by evaporating the chamber acid in leaden boilers until its further concentration would dissolve the lead; it is then transferred to expensive platinum stills, where the evaporation is terminated.

110. The sulphuric acid of commerce always contains a little

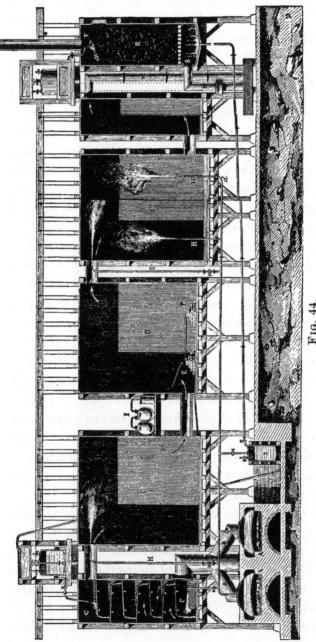

FIG. 44.

lead sulphate, formed in the chambers and evaporating boilers, and when it is diluted with water this lead sulphate becomes in. soluble and separates as a white precipitate. It is often brown from the presence of a little carbonaceous matter. Sometimes the sulphur dioxide is obtained by burning iron pyrites (iron disul- phide), and, as the pyrites often contains arsenic, the resulting sul- phuric acid also contains arsenic. Pure sulphuric acid is made by distilling the commercial acid in glass retorts; the operation requires great care, for the retorts sometimes break, and the vapors of the sulphuric acid are most corrosive and suffocating.

111. *Properties.*—Pure sulphuric acid is a colorless, oily liquid, having at 12° a density of 1.842. It solidifies at 10.5°, and boils at about 338° : its boiling is accompanied by explosive emission of vapor, which may be obviated by putting some pieces of platinum in the vessel. Sulphuric acid is soluble in all proportions of water, and the mixture is accompanied by the production of great heat, showing that there is a true chemical combination between the water and acid. In diluting sulphuric acid with water, we always pour the acid very gradually into the water, which we stir con- stantly. If the mixture is made suddenly, part of the acid is sometimes thrown out of the vessel.

The affinity of sulphuric acid for water is so strong, that the acid causes the formation of water in many substances which do not contain water, but contain hydrogen and oxygen in the pro- portions required for its formation. In a beaker glass, or other thin glass vessel, we pour a little strong solution of sugar, and then some concentrated sulphuric acid : instantly the mixture turns black, and a mass of porous charcoal fills the vessel, which may overflow if we have used too much of the materials. Sugar con- tains carbon or charcoal, and oxygen and hydrogen in the propor- tions to form water. For the same reason a chip of wood with which we stir some sulphuric acid quickly becomes blackened, and the brown color which the acid acquires shows us why the common acid is often brown.

When sulphuric acid is passed through a red-hot tube, it is decomposed into sulphur dioxide, oxygen, and water.

$$H^2SO^4 \quad = \quad SO^2 \quad + \quad O \quad + \quad H^2O$$

We have already seen how zinc acts on sulphuric acid, replacing the hydrogen which then becomes free. The action of copper on the acid is a reducing action, part of the sulphuric acid being reduced to sulphur dioxide.

112. *Molecular structure of sulphuric acid.*—We have seen that hypochlorous acid contains the group hydroxyl, OH; sulphuric acid also contains this group, and we may understand the structure of its molecule by studying some simple reactions. Since a molecule of sulphur dioxide contains two atoms of oxygen, each of which is diatomic,—that is, capable of combining with two atoms of hydrogen,—the sulphur atom must in this compound have as much combining power as four atoms of hydrogen : we call it tetratomic. Yet this sulphur atom is capable of combining with another atom of oxygen; it is *unsaturated* with oxygen, although it is *satisfied* with the two atoms. We mix in a glass jar equal volumes of chlorine and sulphur dioxide, and expose the mixture to direct sunlight: the gases combine to form a colorless liquid, having a suffocating vapor, and we call the compound *sulphuryl chloride.* Analysis shows that it contains SO^2Cl^2, and for convenience' sake the group of atoms SO^2 is called *sulphuryl.* Each atom of chlorine is worth one of hydrogen, and if in sulphur dioxide the sulphur atom is tetratomic, it must be hexatomic in sulphuryl chloride. We may represent this relative combining capacity by little lines, which will show us, not how and where the atoms unite, but the relative worth of the atoms in combination. The free atom of hydrogen or of chlorine would be indicated to be monatomic by a single line, thus, H–, Cl–; and in the molecules of these two elements or of their one compound a single line between the two symbols would show that one has as much combining power as the other :

H–H   Cl–Cl   H–Cl

The symbol of a diatomic element must have two lines, to show that it is worth two monatomic atoms, and we may write water and hydrogen sulphide,

H–O–H   H–S–H

For sulphur dioxide, sulphur trioxide, and sulphuryl chloride, in the first of which the sulphur atom is tetratomic and in the other two hexatomic, we must show the combining power, or *atomicity,* as it is often called, of the elements by four or six lines, and show that the oxygen is diatomic by giving its atoms each two lines. We therefore write,

$$O=S=O \qquad\qquad \begin{matrix} Cl \\ | \\ O=S=O \\ | \\ Cl \end{matrix} \qquad\qquad \begin{matrix} O \quad O \\ \backslash\backslash \, S \, // \\ || \\ O \end{matrix}$$

Sulphur dioxide.   Sulphuryl chloride.   Sulphur trioxide.

When sulphuryl chloride is poured into water, both substances are decomposed, sulphuric and hydrochloric acids being formed.

$$SO^2Cl^2 \quad + \quad 2H^2O \quad = \quad H^2SO^4 \quad + \quad 2HCl$$

We must explain this reaction by a replacement of the chlorine atoms in the sulphuryl chloride by other atoms, or groups of atoms, which have the same combining power, and we would then conclude that sulphuric acid contains two hydroxyl groups, and we might call it *sulphuryl hydrate.*

$$Cl-\overset{O}{\underset{O}{S}}-Cl \ + \ H-O-H \ + \ H-O-H \ = \ H-O-\overset{O}{\underset{O}{S}}-O-H \ + \ HCl \ + \ HCl$$

Sulphuryl   Water (two molecules).   Sulphuric acid.   Hydrochloric acid
chloride.   (2 molecules.)

Analogous study of the manner in which compounds are formed and decomposed, and of the relative worth of the atoms, has led chemists to hold definite ideas of the relations which the atoms bear to each other in a great number of molecules. The group of atoms OH is called a *compound radical* because it is capable of replacing an atom or simple radical. In the same manner the group SO^2 is a radical, as in general are all groups of atoms which pass by double decomposition and without change from one molecule to another. Some radicals, like SO^2, can be separated and studied, because the combining power of the atoms in them, though not saturated, is satisfied; others, like –OH, cannot be separated, because their atoms are not satisfied with each other. Why this is, chemists have not yet been able to explain satisfactorily, but the fact may be stated that a monatomic radical cannot usually exist except in combination: this applies to monatomic atoms as well as to monatomic compound radicals; we have already seen that the molecules of hydrogen and chlorine must each contain two atoms.

There are two other elements whose atoms exactly resemble sulphur in their power of combining with other atoms. They are *selenium* and *tellurium.* They are found only in small quantities, usually associated with gold and silver ores.

LESSON XIV.

SULPHATES.

113. When the hydrogen of sulphuric acid is replaced by metals, sulphates are formed, but as there are two atoms of hydrogen, and either one or both may be replaced, we can understand that there may be two kinds of sulphates. If only one hydrogen atom be replaced, the resulting salt will have acid properties, for it still contains an atom of replaceable hydrogen ; but if both be replaced, we have a *neutral* salt,—that is, one which is neither acid

nor alkaline.   We may study the formation of two of these salts
in the reaction of one and two molecules of sodium hydrate with
one molecule of sulphuric acid.

$$H^2SO^4 \quad + \quad NaOH \quad = \quad NaHSO^4 \quad + \quad H^2O$$
$$\text{Sodium hydrate.} \qquad \text{Sodium acid sulphate.}$$

$$H^2SO^4 \quad + \quad 2NaOH \quad = \quad Na^2SO^4 \quad + \quad 2H^2O$$
$$\text{Sodium sulphate.}$$

But in the action of zinc on sulphuric acid, one atom of zinc
replaces two atoms of hydrogen.   In the same manner, if we boil
sulphuric acid with lead oxide, we have formed lead sulphate, in
which one atom of lead replaces both atoms of hydrogen.

$$PbO \quad + \quad H^2SO^4 \quad = \quad PbSO^4 \quad + \quad H^2O$$
$$\text{Lead oxide.} \qquad\qquad \text{Lead sulphate.}$$

Since one atom of zinc or one of lead is thus capable of replacing
two atoms of hydrogen, those metals are said to be diatomic; and
since sulphuric acid contains two atoms of hydrogen which may be
replaced by two atoms of a monatomic metal, like sodium, or by one
atom of a diatomic metal, like zinc, it is called a *dibasic acid*, and
is capable of forming *neutral* and *acid salts*.

114. With the exception of the sulphates of barium, strontium,
and lead, all of the sulphates are soluble in water, but calcium
sulphate, silver sulphate, and mercurous sulphate are only slightly
soluble.

115. To a solution of magnesium sulphate we add a few drops
of solution of barium chloride or barium nitrate.   A white cloud
forms; this is insoluble barium sulphate; when it has settled, we
may pour off most of the liquid, and we will find that our white
substance is not dissolved by boiling nitric acid.   This test enables
us to recognize either a soluble sulphate or uncombined sulphuric
acid.

Some of the sulphates form anhydrous crystals,—that is, without water; others require water of crystallization.

116. **Sodium Sulphate,** Na^2SO^4, was for a long time called
Glaüber's salt, because Glauber found that it was useful as a purgative medicine.   It crystallizes from water in colorless, rhombic
prisms containing ten molecules of water of crystallization, so that

the formula of the crystals is $Na^2SO^4 + 10H^2O$. They are sol_uble in about ten times their weight of water at $0°$, and in one_third their weight at $33°$; if a saturated solution be made at the latter temperature and immediately sealed, it will remain liquid indefinitely, but on opening the flask the whole of the liquid instantly becomes a mass of crystals.

117. **Potassium Sulphate,** K^2SO^4, forms very hard, colorless crystals, not very soluble in water; it is poisonous.

118. **Calcium Sulphate,** $CaSO^4$.—We have in a beaker glass a very strong solution of calcium chloride. To this we add at arm's length about half its volume of concentrated sulphuric acid: the contents of the beaker at once become so solid that we can invert it and nothing runs out. This solid is calcium sulphate.

$$CaCl^2 + H^2SO^4 = CaSO^4 + 2HCl$$
$$\text{Calcium chloride.} \qquad \text{Calcium sulphate.}$$

Calcium sulphate is the mineral *gypsum, alabaster*, or *selenite*. In these minerals it is combined with two molecules of water of crystallization; this water is driven out when they are heated to $120°$, leaving the anhydrous sulphate as a fine white powder, known as *plaster of Paris*. Unless it has been heated to too high a temperature, this substance will again combine with its water of crystallization, and such combination takes place when plaster of Paris is mixed with water. Plaster casts are made by mixing the plaster and water to a creamy consistence, and pouring the liquid into the moulds: in a few minutes the plaster *sets*, or becomes hardened, and in so doing it expands and completely fills the mould. Calcium sulphate dissolves in about 500 times its weight of water. It is a valuable fertilizer for certain soils.

119. **Strontium Sulphate,** $SrSO^4$, constitutes the mineral *celestine*, so called because it often has a blue color, though the pure salt is white. It is insoluble in water, and is precipitated when a soluble strontium salt is added to sulphuric acid or a soluble sulphate.

120. **Barium Sulphate,** $BaSO^4$, is found native as *heavy spar*. We have seen that it is formed by the reaction of sulphuric acid with soluble salts of barium. It is sometimes used for adulterating white lead (§ 250).

121. Magnesium Sulphate, $MgSO^4 + 7H^2O$, is commonly known as Epsom salts. It is made by dissolving magnesium carbonate in dilute sulphuric acid, and when the concentrated solution is allowed to evaporate, the salt separates in crystals containing seven molecules of water. It has a salty, bitter, and unpleasant taste. It dissolves in about three times its weight of water. It is used in medicine.

122. Zinc Sulphate, $ZnSO^4 + 7H^2O$.—We evaporate to a small volume the liquid remaining in the bottle in which we made hydrogen by the action of sulphuric acid on zinc, and then set it aside in a cool place. After a time zinc sulphate separates in beautiful transparent crystals, containing seven molecules of water, and of exactly the same form as those of magnesium sulphate prepared in the same manner. Compounds which have in their molecules the same number of atoms arranged in the same manner, usually crystallize in the same form, and are said to be *isomorphous*. Zinc sulphate is sometimes called white vitriol. It is quite soluble in water, and when swallowed it acts as a violent emetic.

123. Ferrous Sulphate, $FeSO^4 + 7H^2O$.—This salt, called also green vitriol and copperas, is made by treating scrap iron with dilute sulphuric acid; hydrogen is disengaged, just as in the action of the same acid on zinc. When the filtered solution is evaporated and set aside to crystallize, the ferrous sulphate separates in crystals which, as the formula would indicate, are isomorphous with the two preceding salts. These crystals are pale green in color; when exposed to dry air, they lose part of their water of crystallization, and the surface becomes covered with a white powder, which is the anhydrous salt; they are said to *effloresce*. After a time this powder becomes yellow, from an absorption of oxygen (§ 527). Ferrous sulphate is soluble in less than twice its weight of cold water, and much more soluble in boiling water. It is poisonous.

124. Cupric Sulphate, $CuSO^4 + 5H^2O$.—This beautiful blue salt, often called blue vitriol, may be prepared from the residue of the preparation of sulphur dioxide by diluting it with water, filtering, and evaporating to crystallization. It is usually made by *roasting*—that is, heating in the air—copper sulphide (§ 484), and treating the mass with water. When the blue crystals are heated, the water is driven out, and the white anhydrous salt is left. Cupric sulphate dissolves in four times its weight of cold, or twice its weight of boiling water. To a solution of this salt we add a little ammonia water; a pale-blue precipitate forms, but when we add more ammonia this precipitate again dissolves, and a deep-blue liquid is obtained. This liquid contains *ammoniacal cupric sulphate*. Cupric sulphate is used in telegraphic batteries, in dyeing, for electrotyping, and in many other operations. It is poisonous.

125. Lead Sulphate, $PbSO^4$.—When sulphuric acid or the solution of a sulphate is added to a solution containing a lead salt, lead sulphate separates as a white precipitate. It occurs in nature as the mineral *anglesite*. It is insoluble in water, but dissolves in strong acids.

Of the many other sulphates, we must study a few when we shall have learned some of the peculiarities of the corresponding metals.

LESSON XV.

NITROGEN.—THE ATMOSPHERE.

126. Nitrogen, $N = 14$.—On the water in the pneumatic trough, we float a small capsule containing a little sand on which we have placed a piece of phosphorus. We ignite the phosphorus, and place over it a bell-jar which may rest on the shelf in the trough (Fig. 45). At first, as the heat of the burning phosphorus expands the air, a few bubbles of air escape under the edge of the jar, but this soon stops; presently the water begins to rise in the jar, and the phosphorus no longer burns. All the oxygen of the air in the jar has been consumed by the phosphorus, and there is left only nitrogen, with which in the free state the oxygen was mixed, and phosphoric oxide. The latter will presently dissolve in the water, and we may then examine the nitrogen.

To obtain nitrogen from the air, it is only necessary to remove

the oxygen. When we wish a considerable quantity of pure nitrogen, we force a current of air through a tube containing pieces of solid potassium hydrate, which absorbs the moisture and carbon dioxide, and from this through a long tube containing red-hot copper. The copper combines with all of the oxygen, forming cupric oxide, and pure nitrogen passes out at the end of the tube.

FIG. 45.

127. Nitrogen is a colorless, tasteless, and odorless gas. Its density compared to air is 0.97, or compared to hydrogen, 14 ; as its atomic weight is also 14, its molecule must contain two atoms. It is almost insoluble in water. It is not combustible, neither will it support the combustion of other substances. It combines directly with only a few of the elements, and energy is absorbed during the formation of many of its compounds; that is, the nitrogen atoms have a stronger affinity for one another than for the other atoms with which they are combined.

Before considering any of these compounds, we must study the composition of the atmosphere from which the nitrogen is derived.

128. **The Atmosphere.**—The chemical composition of the air was first determined with tolerable accuracy by the great French chemist, Lavoisier. We may satisfy ourselves of this composition in a very simple manner. Around a long glass tube, closed at one end, we have placed four caoutchouc bands, dividing it into five equal portions. Into this tube, which must be perfectly dry, we drop a dry piece of phosphorus, and tightly cork the open end. By gently heating the bottom of the tube over a lamp, we inflame the phosphorus, and then by quickly turning the tube

bottom up and giving with the corked end a few sharp blows on the table, we cause the burning phosphorus to fall the whole length of the tube. If our experiment has been well made, all the oxygen has been burned from the air in the tube, which we allow to cool, and then carefully uncork with the mouth under̄water. As soon as the cork is drawn, the water rises to the first division (Fig. 46). The air which we have roughly analyzed, then, contained about one-fifth oxygen and four-fifths nitrogen by volume.

FIG. 46.

129. A very accurate analysis of air is made by the aid of the eudiometer, which we have studied. Into the eudiometer with the caoutchouc tube and plain glass tube, which served for the synthesis of water (§ 42), but without the enclosing wide glass tube, we introduce 100 measures of air and 100 measures of pure hydrogen. After adjusting the mercury level in the two tubes, we pass an electric spark: at once the oxygen and part of the hydrogen are converted into water, which condenses, and the volume of gas is re--duced. We know that water is formed by the union of two volumes of hydrogen and one volume of oxygen ; consequently one-third of the diminution in volume must be caused by the removal of the oxygen of the 100 measures of air. On again adjusting the level of the mercury, we find that instead of 200 measures we have only 137.21. The oxygen present in 100 measures of air must, then, have been $\dfrac{200 - 137.21}{3}$, or 20.93 measures. We conclude, therefore, that 100 volumes of air contain 20.93 volumes of oxygen and 79.07 volumes of nitrogen.

130. Since oxygen is heavier than nitrogen, these relative volumes will not express the relations by weight. We can calculate the weights from the volumes, and the result would show us that 76.87 parts by weight of nitrogen are mixed with 23.13 parts of

oxygen. These proportions are confirmed by the direct analysis, which is made by passing air through a series of tubes in which all traces of everything but oxygen and nitrogen are absorbed: thus purified, the air passes through a tube containing red-hot copper (§ 127), and the increase in weight of this tube gives the amount of oxygen in the air analyzed. The nitrogen passes on into a glass globe in which a vacuum has previously been made, and of course the increased weight of this globe is the amount of nitrogen.

The air is not a compound, but a mixture, and we may expect that the proportions of the constituents shall vary a little. However, the composition is about constant: hundreds of analyses have shown that the proportion of oxygen in 100 volumes of unconfined air varies only from 20.9 to 21. Over large surfaces of water, as on the open sea, there is a little less oxygen (20.6), because that gas is more soluble in water than is the nitrogen.

131. We pour into a plate some perfectly clear lime-water; in a few minutes a thin, white pellicle forms over its surface. This pellicle is calcium carbonate, and has been formed by the absorption of carbon dioxide from the air. Air also contains more or less vapor of water, which is deposited in the form of dew on very cold objects.

The proportions of vapor of water and carbon dioxide may be determined by drawing a known volume of air through a series of tubes (Fig. 47), the first of which contain pumice-stone and sulphuric acid, and the others fragments of potassium hydrate. The sulphuric acid absorbs the vapor of water, and the increase in weight of the first tubes (D, E, F) gives the weight of that vapor. The carbon dioxide is absorbed by the potassium hydrate, and the increase in weight of the tubes containing it (A, B, C) gives us the proportion of that gas. The volume of air which contained these quantities of carbon dioxide and watery vapor is equal to the volume of water which runs from the aspirator (V), the air passing through the tubes and into the aspirator to take the place of the water running out. We can calculate the weight of this air from its volume, for at 0° and under 760 millimetres barometric pressure, one litre of dry air weighs 1.2932 grammes.

132. The quantity of vapor of water which the air can take up depends on the temperature, and air is said to be saturated with moisture when at the given temperature it can hold no more

water vapor. It is then said to have a *relative humidity* of 100: at the same temperature half that quantity of vapor would be a relative humidity of 50. But if the temperature be increased

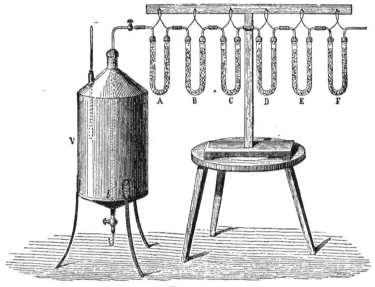

FIG. 47.

and the quantity of moisture remain the same, the relative humidity is lowered, for the air is then capable of dissolving more vapor. The temperature at which air is completely saturated with vapor is called the *dew-point*, and this may be determined by noting the temperature at which moisture begins to deposit on the walls of a vessel which is artificially cooled.

Substances which are capable of absorbing the moisture from the atmosphere are said to be hygroscopic. By reason of its affinity for water, sulphuric acid placed in an open vessel will in a few days absorb from the air enough moisture to double its volume.

133. Carbonic acid gas is present in the air in only small proportions; from four to six parts in ten thousand parts of air. It is thrown into the atmosphere from volcanoes, fissures in the earth, and mineral springs, but the largest quantity is produced by combustion and respiration. It does not accumulate in the atmosphere,

but is absorbed by plants, and under the influence of sunlight is decomposed, the carbon being retained for the growth of the plant,

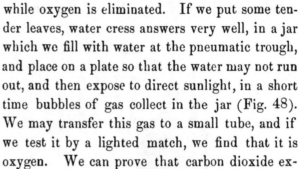

while oxygen is eliminated. If we put some tender leaves, water cress answers very well, in a jar which we fill with water at the pneumatic trough, and place on a plate so that the water may not run out, and then expose to direct sunlight, in a short time bubbles of gas collect in the jar (Fig. 48). We may transfer this gas to a small tube, and if we test it by a lighted match, we find that it is oxygen. We can prove that carbon dioxide exists in the air exhaled from the lungs, by blowing the breath through lime-water (Fig. 49), which quickly becomes clouded by the formation of calcium carbonate. In the same manner, if we burn a

FIG. 48.

lighted taper or candle in a covered jar, and then pour in some lime-water, and shake the jar, the milkiness of the water shows that carbon dioxide has been formed.

134. Although in unconfined air, plants and vegetables remove the carbon dioxide, so that its proportion does not increase, yet if the air be confined, as in a room or a mine, this gas may accumulate to as much as one part in a hundred of air. As this carbon dioxide is formed at the expense of the oxygen of the air, the proportion of oxygen may descend as low as 22 parts per hundred by weight, instead of 23.13. A single gas-burner burning about 160 litres of gas per hour, consumes the oxygen of about 240 litres of air, and produces about 130 litres of carbon dioxide. At every breath, a man consumes about 4.87 per cent. of the oxygen which he inhales, and the

FIG. 49.

carbon dioxide exhaled in an hour is about 20 litres. When the carbon dioxide in the air is pure, its proportion may be much increased, and no ill effects result; but in addition to this gas a considerable proportion of animal matters passes from the lungs, and, together with that thrown off in the perspiration, quickly vitiates the atmosphere of an apartment which is not ventilated. Good ventilation requires from six to ten thousand litres of air per hour for each individual. In dwellings and workshops, most of the ventilation is by the cracks of doors and windows. Fires in open grates afford excellent ventilation, the draught of the chimney drawing a constant supply of air into the room.

135. Besides the substances already considered, air always contains very small quantities of ammonia, traces of nitric acid, and small solid particles of various natures which are carried to great distances by the winds. Sometimes a little ozone is present, and may be recognized by the test which we have studied (§ 66).

LESSON XVI.

AMMONIA AND ITS COMPOUNDS.

136. **Ammonia,** NH^3.—In a glass flask to which we have adapted a cork and delivery-tube, we mix some powdered ammonium chloride with its own weight of powdered quick-lime. We then fill the rest of the flask with pieces of lime, and gently heat it on a sand-bath. We soon notice the pungent odor of the gas disengaged; as this gas is very soluble in water, we cannot collect it

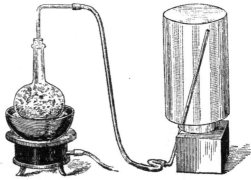

Fig. 50.

over that liquid; we may collect it either over mercury in a small pneumatic trough, or by upward dry displacement, for it is lighter than air (Fig. 50). Quick-lime is calcium oxide, CaO; ammonium chloride is a compound of nitrogen, hydrogen, and chlorine, NH^4Cl. We may write the reaction,

$$2NH^4Cl \quad + \quad CaO \quad = \quad 2NH^3 \quad + \quad CaCl^2 \quad + \quad H^2O$$
Ammonium chloride.　　Lime.　　Ammonia.　　Calcium chloride.

The calcium chloride formed remains in the flask, and the water is absorbed by the pieces of lime which we have put into the flask for that purpose. We could not dry ammonia gas by passing it over either calcium chloride or sulphuric acid, for it combines with both of those substances.

137. Properties.—The ammonia which we have collected is a colorless gas, having a penetrating, pungent odor, and a burning taste. We must not inhale too much of it, for, although not poisonous, it often produces sudden giddiness or vertigo. Its density compared to hydrogen corresponds with half its molecular weight, being 8.50: it is therefore a little more than half as heavy as air. By strong pressure, it is readily converted into a liquid, and this liquid is employed in some forms of ice-machines, where it produces great cold by its evaporation.

FIG. 51.

Ammonia is very soluble in water: at 0° water will dissolve 1000 times its volume of the gas, and at ordinary temperatures about 700 times its volume. We have fitted to a glass flask a cork through which passes a tube drawn out to a small opening on the inside. We fill this flask with ammonia, by dry displacement, and after putting in the cork we dip the end of the tube into a vessel of water. The water slowly rises in the tube, but as soon as it reaches the narrow end the ammonia is absorbed so rapidly that the pressure of the atmosphere forces the water up in a fountain which continues until all of the ammonia is dissolved. (Fig. 51.) The solution of ammonia in water is called ammonia-water, or, more commonly, spirits of hartshorn. It has the taste and odor of the gas, and is very caustic. When it is heated, the gas is driven out, and we may most readily obtain ammonia by heating strong ammonia-water in a flask, and drying the gas by passing it through a tube containing quick-lime. The strong ammonia-water of commerce contains about 35 per cent. of the gas. Its density is about 0.86.

138. Ammonia is decomposed into nitrogen and hydrogen by very high temperatures or by the continued passage of electric sparks. Two volumes of ammonia yield four volumes of the mixed gases, and if we mix in the eudiometer these four volumes with one and a half volumes of oxygen and pass the spark, after the condensation of the water formed, only one volume of gas is left. This is nitrogen, and two volumes, or one molecule, of ammonia must therefore contain one volume (one atom) of nitrogen, and three volumes (three atoms) of hydrogen.

139. Ammonia is combustible, but it will not burn in the air. We may cause it to burn at a jet which is surrounded by oxygen, and for that purpose we have fitted to a short wide tube, open at both ends, a cork through which pass two tubes (Fig. 52); one of them is short and leads oxygen from á gas-holder, while the other reaches nearly to the top of the wide tube, and conveys ammonia gas from a small flask in which we boil some ammonia-water. As soon as ammonia-gas escapes from the jet, we turn on the oxygen, and light the ammonia, which burns with a yellow flame, forming water and nitrogen.

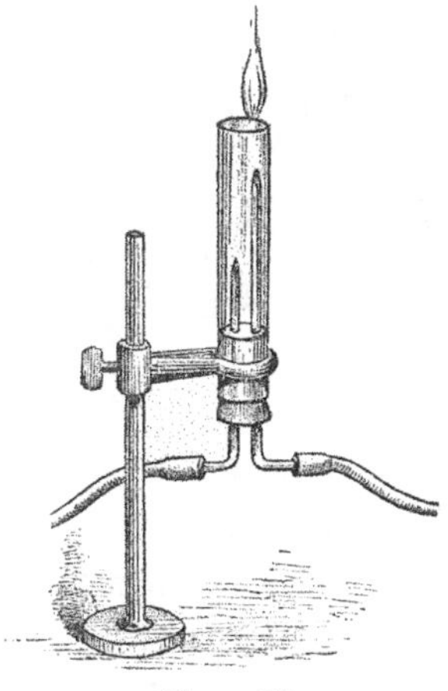

$$4NH^3 + 3O^2 = 6H^2O + 2N^2$$

FIG. 52.

This combustion may be made to take place more slowly, and in an interesting manner, in the presence of platinum. Over some ammonia-water contained in a beaker glass (Fig. 53) we suspend a coil of red-hot platinum wire, so

that it may nearly touch the liquid.   The coil will continue to
glow for a long time by the heat evolved from the slow combustion

of the ammonia which escapes from the liquid
and mixes with the oxygen of the air.   If now
we warm the beaker, and pass bubbles of oxygen
through the liquid, each bubble causes a little
explosion as it combines with the hydrogen of
the ammonia.   Sometimes the beaker becomes
filled with white fumes of a compound known
as ammonium nitrite.

FIG. 53.

140. Ammonium Compounds.—Ammonia is the only compound of nitrogen and hydrogen which chemists have been able
to prepare, and it seems that one atom of nitrogen has affinities
for only three atoms of hydrogen.   However, that same atom
may combine with another atom of hydrogen if certain other elements be present.

Over a small capsule containing some warm ammonia-water we
have inverted a glass jar, and, at a little distance, over another cap-

sule in which is some
warm   hydrochloric
acid we have inverted
another jar.   Each
jar now contains some
of the gas from the
liquid under it.   When
we raise the jars and
bring their mouths to-

FIG. 54.

gether, both become filled with dense white fumes (Fig. 54).
The two gases have combined, and a body called ammonium chloride has been formed, and will settle on the sides of the jars.   The
combination is very simply expressed.

$$NH^3 + HCl = NH^4Cl, \text{ ammonium chloride.}$$

We see, then, that while the nitrogen atom will combine with
only three atoms of hydrogen alone, it will combine with four if
an atom of chlorine come with that hydrogen.   In the same manner, in many other compounds one nitrogen atom is combined with

four hydrogen atoms, and one other atom or group of atoms. NH^4 is one of those groups of atoms which we call radicals (§ 112); it passes from one compound to another without change, just as an atom of hydrogen may pass from one molecule to another. It cannot, however, be separated in the free state from any of these compounds. It is called ammonium.

LESSON XVII.

AMMONIUM COMPOUNDS.

141. Ammonium Chloride, NH^4Cl.—This compound is formed by the direct union of ammonia and hydrochloric acid. During the manufacture of illuminating gas by the distillation of coal, more or less ammonia is formed; it must be removed before the gas is fit for use, and this is accomplished by washing the gas with water (§ 225). A dilute solution of ammonia is thus obtained, and this is the source of the ammonia and ammonium compounds of commerce. For the preparation of ammonium chloride this *gas liquor* is heated with lime, and the ammonia gas given off is passed into hydrochloric acid. The solution is then evaporated, and the residue of ammonium chloride is purified by sublimation in stoneware pots. It may be formed by another and interesting reaction : we pass into a jar of dry chlorine the drawn-out end of a tube through which ammonia is escaping; at once the ammonia takes fire,

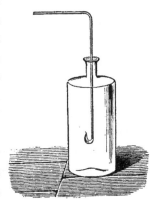

Fig. 55.

being partially decomposed with production of hydrochloric acid, which at once unites with another portion of the ammonia, forming white clouds of ammonium chloride (Fig. 55).

$$2NH^3 + 3Cl^2 = N^2 + 6HCl$$
$$6HCl + 6NH^3 = 6NH^4.Cl$$

When pure, ammonium chloride is in translucid masses, which have a fibrous structure, and are quite tough and difficult to pulverize. It dissolves in two and a half times its weight of cold water, and in much less hot water. Its taste is not unpleasantly salty and sharp. Unless in large doses, it is not poisonous.

142. **Ammonium Sulphate,** $(NH^4)^2SO^4$, is manufactured by passing into dilute sulphuric acid the ammonia which is disengaged when gas liquor is heated with lime. It is in white, colorless crystals, readily soluble in water, having a sharp taste. It may be used for the manufacture of ammonia, and is employed as a fertilizer.

143. **Ammonium Sulphydrate,** $NH^4.SH$.—We have already noticed the composition and mode of formation of potassium sulphydrate (§ 101). When hydrogen sulphide is passed into ammonia-water until the liquid will dissolve no more of the gas, ammonium sulphydrate is formed.

$$NH^3 \quad + \quad HSH \quad = \quad NH^4.SH$$

It is a colorless liquid, but becomes yellow after it has been for some time exposed to the air. Its odor is disgusting, being at the same time that of hydrogen sulphide and that of ammonia. If it be mixed with a quantity of ammonia-water exactly equal to that from which it was prepared, ammonium sulphide is formed.

$$NH^4SH \quad + \quad NH^3 \quad = \quad NH^4.S.NH^4 \quad = \quad (NH^4)^2S$$
Ammonium sulphide.

This compound is of much value in the laboratory in detecting some metals.

To a solution of ferrous sulphide we add a few drops of ammonium sulphide, and a black precipitate of ferrous sulphide is thrown down.

$$NH^4.S.NH^4 \quad + \quad FeSO^4 \quad = \quad (NH^4)^2SO^4 \quad + \quad FeS$$
Ammonium sulphide. Ferrous sulphate. Ammonium sulphate. Ferrous sulphide.

We pour a few drops of the same liquid into a solution of zinc sulphate, and white zinc sulphide is precipitated.

$$(NH^4)^2S \quad + \quad ZnSO^4 \quad = \quad (NH^4)^2SO^4 \quad + \quad ZnS$$

144. On examining the composition of the ammonium compounds, we see that the radical NH^4 has the same combining power as one atom of hydrogen.

It is a monatomic radical; but at the same time we notice that it can replace the hydrogen atoms in the acids, and in so doing it forms salts. It is a *basic radical*, and is in this respect exactly opposite to the radicals ClO_ and SO², which are *acid radicals*.

145. Ammonium Amalgam.—We make an amalgam of sodium,—that is, a compound of sodium and mercury,—by throwing on the surface of a little mercury a few small pieces of sodium. If these do not at once combine with the mercury, we can readily effect the combination by touching them with a drop of water on the end of a long glass rod. As little pieces of burning sodium are sometimes thrown out, we keep the vessel at a sufficient distance from the eyes. We now pour this amalgam into a tall jar containing a strong solution of ammonium chloride: at once a very curious phenomenon occurs. The mercury begins to swell and become pasty; it rises and floats on the water, and sometimes it overflows the jar. On pouring it out and examining it, we find that it has become a brilliant, butter-like substance, very light, and entirely unlike the mercury. In the reaction which has taken place, the sodium has left the mercury and combined with the chlorine of the ammonium chloride. The ammonium radicals thus set free have at once combined with the mercury, forming this body which we call the ammonium amalgam.

$$2NH^4.Cl \quad + \quad HgNa^2 \quad = \quad 2NaCl \quad + \quad NH^4.Hg.NH^4$$
Ammonium chloride. Sodium amalgam. Sodium chloride. Ammonium amalgam.

Ammonium amalgam will amalgamate with iron, a property which ordinary mercury does not possess. It does not keep long, but soon decomposes into mercury, ammonia gas, and hydrogen.

146. Nitrogen Iodide.—We have reduced a small quantity of iodine to a fine powder, and we throw this into a little ammonia-water. Part of it dissolves, and the other part is converted into a black powder, which we carefully pour on a small filter placed in a funnel. When most of the liquid has drained off, we distribute this powder on several pieces of filter-paper, which we set aside for the powder to dry. When it is dry, the lightest touch causes it to explode with a loud noise, and sometimes it explodes spontaneously. In any case the explosion is always accompanied by the production of purple vapors of iodine. The black powder is nitrogen iodide: there are several such compounds, and their composition depends on the exact manner of formation; we may express it by NI^3. It is formed by a reaction which yields also ammonium iodide.

$$4NH^3 \;+\; 3I^2 \;=\; 3NH^4I \;+\; NI^3$$
Ammonia.   Iodine.   Ammonium iodide.   Nitrogen iodide.

The ammonium iodide is formed with a considerable production of energy; but the liquid does not become warm, for all this energy is transferred to the nitrogen and iodine atoms which combine to form nitrogen iodide. Where must we seek the energy of explosion of nitrogen iodide? The explosion is only a rearrangement of the atoms; a decomposition of the nitrogen iodide; the energy of this decomposition we must consider as the energy of formation of nitrogen molecules and iodine molecules, of which the atoms then disengage the energy conferred on them by the formation of ammonium iodide, and retained in the nitrogen iodide.

LESSON XVIII.

OXIDES OF NITROGEN.

147. Nitrogen Monoxide, N^2O.—In a glass flask on a sand-bath, we heat some ammonium nitrate, a white, crystalline substance obtained by neutralizing ammonia with nitric acid. Our flask being provided with a delivery-tube, we may collect the gas over the pneumatic trough (Fig. 56). The ammonium nitrate is entirely decomposed into water and nitrogen monoxide sometimes called nitrous oxide.

$$NH^4NO^3 \;=\; N^2O \;+\; 2H^2O$$
Ammonium nitrate.   Nitrogen monoxide.

As the water is converted into steam by the heat required for the experiment, when we desire to collect the gas in a gas-bag we pass it first through an empty bottle in which the steam may condense.

148. Nitrogen monoxide is a colorless gas, having no odor, but a sweet taste. Its density is 22 compared to hydrogen, or 1.527 compared to air. It is liquefied by great pressure, and considerable quantities are so liquefied in strong iron cylinders, in

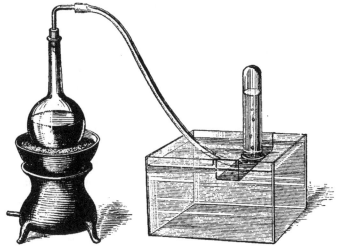

Fig. 56.

order that the gas may be transported in small bulk for the use of dentists. At ordinary temperatures, water dissolves about its own volume of nitrogen monoxide; for this reason some of the gas is always lost when it is collected over water.

Nitrogen monoxide is decomposed by heat, two volumes of the gas yielding two volumes of nitrogen and one of oxygen. Since the gaseous mixture contains a much larger proportion of oxygen than does the air, nitrogen monoxide should support combustion better than the air. An experiment will show us that it does; we put into a jar of nitrogen monoxide gas a taper bearing only a spark of fire, and this spark, decomposing the gas surrounding it, sets free sufficient oxygen to relight the taper (Fig. 57). In the same manner phosphorus and sulphur burn brilliantly in this gas.

Nitrogen monoxide is not poisonous; it may be inhaled for a short time without danger, and its inhalation is followed by insensibility, a condition called anæsthesia. Advantage is taken of this property of the gas for the performance of short surgical operations. The first effects of the inhalation of the gas are often a condition of excitement and disposition to gayety; for this reason it has been called laughing-gas.

Fig. 57.

149. Nitrogen Dioxide, NO.—In a gas-bottle, provided with a delivery-tube and funnel-tube, we have some copper clippings and water. Through the funnel-tube we pour nitric acid until there is a brisk disengagement of gas. At first this gas in the gas-bottle is red, for reasons which we shall presently learn, but soon it becomes almost colorless. We then pass the delivery-tube under water, and collect the gas in jars filled with water (Fig. 58). In the reaction which is taking place, the copper is replacing the hydrogen of the nitric acid, and every atom of copper replaces two atoms of hydrogen.

$$Cu \quad + \quad 2HNO^3 \quad = \quad Cu(NO^3)^2 \quad + \quad 2H$$
Copper.   Nitric acid.   Cupric nitrate.

But in this case the hydrogen is not set free; it reduces more nitric acid, and if we keep our generating bottle cool by placing it in cold water, the reduction yields NO and water. As the copper atoms always set free even numbers of hydrogen atoms, we cannot write this reaction $3H + HNO^3 = 2H^2O + NO$, but must write $6H + 2HNO^3 = 4H^2O + 2NO$; and since the six atoms of hydrogen must be replaced by three atoms of copper, each of which requires two molecules of nitric acid besides the two that are reduced, we may write the whole equation

$$3Cu + 8HNO^3 = 3Cu(NO^3)^2 + 4H^2O + 2NO$$

While this gas is called nitrogen dioxide, we must remember that its molecule does not contain two atoms of oxygen : the name will only help us to remember that it is the second oxide of nitrogen. A better name is nitric oxide.

150. Nitric oxide is a colorless gas, of which we must remain ignorant of the taste and odor, for it forms a corrosive gas as soon as it is exposed to the air. Its density compared to air is 1.039. It is almost insoluble in water. It has been liquefied by great cold and pressure.

It is decomposed by heat, but not so readily as nitrogen monoxide. For this reason, al-

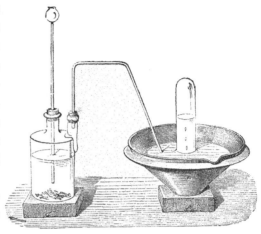

FIG. 58.

though it contains in a given volume twice the proportion of oxygen in nitrogen monoxide, it will not relight a taper bearing a spark: it will, however, support the combustion of phosphorus and charcoal.

The most remarkable property of nitric oxide is its affinity for oxygen. We uncover a jar filled with the gas, and instantly a cloud of red vapor is formed. This is the red gas which was formed in the generating bottle when the nitric oxide first elimi-nated came in contact with the air in the bottle. In this experi-ment one molecule of nitrogen dioxide takes an atom of oxygen from the air, and the red vapor is the gas NO^2. We must be careful not to inhale this gas, for it is very injurious.

We pour a few drops of carbon disulphide into a jar of nitric oxide. The vapor of this volatile liquid at once mixes with the gas, and when we apply a flame, a bright flash of light fills the jar as the carbon is burned by the oxygen of the nitric oxide. The light produced by this little explosion affords an excellent means for causing the direct combination of hydrogen and chlorine (§ 71).

We pour a little ferrous sulphate solution into a jar of nitric oxide; some of the gas is at once absorbed, and the liquid becomes brown. The nitric oxide may be driven out by heating the solution, and the pure gas is sometimes prepared in this manner.

151. In nitric oxide the affinities of the nitrogen atom are not exhausted: we have seen that it is still able to combine with an atom of oxygen. It will also combine with an atom of chlorine; when one volume of chlorine is mixed with two volumes of nitric oxide, the gases unite, forming a compound NOCl. When this compound is treated with water, both substances are decomposed, yielding hydrochloric acid and nitrous acid, HNO^2.

$$NOCl + H^2O = HNO^2 + HCl$$

Nitric oxide may, then, act as a radical, and in its compounds it is called *nitrosyl*. NOCl is therefore called *nitrosyl chloride*, and nitrous acid may be called *nitrosyl hydrate*, NO–OH.

LESSON XIX.

OXIDES OF NITROGEN (Continued).

152. **Nitrogen Peroxide,** NO^2 and N^2O^4.—We may form this substance by the direct combination of nitric oxide and pure oxygen, and we would of course require two volumes of the first and one volume of the second. We can prepare it in another manner. We

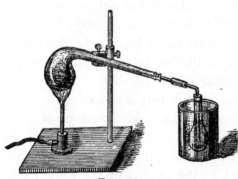

FIG. 59.

heat some dry lead nitrate in a small retort placed in a sand-bath. The vapors given off are conducted into a flask surrounded by ice (Fig. 59). Because the lead nitrate cannot well be perfectly dried, we change the receiver after a little liquid has collected in it, and throw away this first portion. That which now collects has a yellow color, and if we mix a little salt with the ice around the receiver, the liquid will freeze to a crystalline mass. The lead nitrate is decomposed into lead oxide, oxygen, and nitrogen peroxide.

$$Pb(NO^3)^2 \quad = \quad PbO \quad + \quad O \quad + \quad N^2O^4$$

Lead nitrate.   Lead oxide.   Nitrogen peroxide.

The solid nitrogen peroxide melts at —10° to a nearly colorless liquid; this liquid becomes yellow and afterwards orange-colored as the temperature rises, and at 15° is red. It boils at 22°, giving the red vapor, and the density of this vapor compared to hydrogen is 46, showing that the molecular weight of the compound is 92, and the molecule must, therefore, contain N^2O^4. However, as the temperature rises the density diminishes, and at 70° it is only one-half 46; after this, the density remains constant, the molecule has become two molecules, and each of these must contain NO^2. Such decomposition of gases by heat is called dissociation: we have already noticed the dissociation of water vapor (§ 54) and nitrogen monoxide.

153. Nitrogen peroxide dissolves in water, but in dissolving it reacts with the water; with a small quantity of water it forms nitrogen trioxide and nitric acid, while with a larger quantity it yields nitric acid and nitrogen dioxide.

$$2N^2O^4 \quad + \quad H^2O \quad = \quad 2HNO^3 \quad + \quad N^2O^3$$

Nitrogen peroxide.   Nitric acid.   Nitrogen trioxide.

$$3N^2O^4 \quad + \quad 2H^2O \quad = \quad 4HNO^3 \quad + \quad 2NO$$

With the alkaline hydrates it yields nitrates and nitrites.

$$N^2O^4 \quad + \quad 2NaOH \quad = \quad NaNO^3 \quad + \quad NaNO^2 \quad + \quad H^2O$$

Sodium hydrate.   Sodium nitrate.   Sodium nitrite.

A similar decomposition really takes place with water, but the nitrous acid formed is at once decomposed by the water.

The red vapors are dangerous to inhale, and the more dangerous because they do not give immediate discomfort. They act on the delicate membrane of the lungs, and there have been many fatal accidents where the gas has been inhaled by workmen repairing sulphuric acid chambers (§§ 108, 109).

154. Just as nitrogen dioxide may act as a radical which we called nitrosyl, nitrogen peroxide, NO^2, may act as a radical. By means which we need not consider, chlorine may be made to combine with nitrogen peroxide, and the resulting compound, which is called nitryl chloride, is a volatile liquid whose

molecule is expressed by the formula NO^2Cl. When this liquid is poured into water, nitric acid and hydrochloric acid are formed.

$$NO^2Cl + H^2O = HNO^3 + HCl$$

We may therefore regard NO^2 as the radical of NO^2Cl and HNO^3, and as a radical it is called nitryl. NO^2Cl is then *nitryl chloride*, and nitric acid is *nitryl hydrate*, $NO^2.OH$.

155. **Nitrogen Pentoxide,** N^2O^5.—When vapor of nitryl chloride is passed over silver nitrate, heated in a tube to 70°, the silver and chlorine atoms leave their respective molecules and combine together, forming silver chloride; the two groups of atoms from which the chlorine and silver are removed also combine and form a volatile solid compound that condenses in the cooler part of the tube. This body is nitrogen pentoxide, N^2O^5.

$$\underset{\text{Nitryl chloride.}}{NO^2\text{-}Cl} + \underset{\text{Silver nitrate.}}{AgO\text{-}NO^2} = \underset{\text{Silver chloride.}}{AgCl} + \underset{\text{Nitrogen pentoxide.}}{NO^2\text{-}O\text{-}NO^2}$$

It is a colorless body, melting at 29.5°, and boiling at 50°. It is easily decomposed, and sometimes explodes spontaneously.

156. In studying the other elements we have examined the combining powers of their atoms, compared to that of an atom of hydrogen,—that power to which the name atomicity or *valence* (worth) has been given. What is the atomicity of the nitrogen atom? The molecule of nitrogen monoxide closely resembles in structure the molecule of water: replace by two atoms of nitrogen the two atoms of hydrogen of water, and we have a molecule of nitrogen monoxide. The nitrogen atoms here have the same combining power as the hydrogen atoms, and we say they are monatomic. In ammonia, however, the nitrogen atom itself combines with three atoms of hydrogen; it must then be worth three hydrogen atoms, and we call it triatomic. But in ammonium chloride it is united with four hydrogen atoms and one chlorine atom; since we have agreed that the chlorine atom has the same worth as the hydrogen atom, the nitrogen atom in ammonium chloride must be pentatomic. Now let us look at the other oxygen compounds of nitrogen: we have seen that in the compounds already studied the oxygen atom is diatomic, and indeed we shall in time find that there are many reasons for believing that oxygen is always diatomic. Then in nitric oxide, NO, the nitrogen atom, which is combined with only one oxygen atom, must also be diatomic; but we have seen that this compound NO combines directly with a chlorine atom, forming the compound nitrosyl chloride, NO–Cl: it combines with a hydroxyl group, which is monatomic, forming nitrous acid, HO–NO, and in these compounds the nitrogen must be triatomic. We must conclude, however, that in NO^2 the nitrogen is tetratomic, since it is combined with two atoms of diatomic oxygen, but here again an atom of monatomic chlorine will unite with the nitrogen atom which is then pentatomic in nitryl chloride, NO^2Cl. At low temperatures, when the red vapors condense, forming molecules of N^2O^4, it seems also that nitrogen is pentatomic, and that in the molecule of N^2O^4 two nitrogen atoms, each of which is combined with two oxygen atoms, are also combined with each other.

If now we remember our representations of the combining powers of the atoms by short lines, we may see how the atoms in these molecules seem to be related, and how the atomicity of nitrogen varies.

$$\begin{array}{ccccc} & & \overset{\textstyle H}{\underset{\textstyle |}{}} & & \overset{\textstyle Cl}{\underset{\textstyle |}{}} \\ N-O-N & H-N-H & N=O & Cl-N=O & O=N=O \end{array}$$

Nitrogen monoxide.   Ammonia.   Nitrogen dioxide.   Nitrosyl chloride.   Nitryl chloride.

The reactions between nitrosyl chloride and water, and nitryl chloride and water, then become double decompositions which we can easily understand.

$$\overset{\textstyle Cl}{\underset{\textstyle O=N=O}{|}} \; + \; H-O-H \; = \; \overset{\textstyle O-H}{\underset{\textstyle O=N=O}{|}} \; + \; HCl$$

and $O=N-Cl \; + \; H-O-H \; = \; O=N-O-H \; + \; H-Cl$

We can understand also how nitrogen peroxide decomposes by the action of water, yielding nitric and nitrous acids.

$$\begin{array}{ccccc} O=N=O & & & O=N=O & \\ \underset{\textstyle O=N=O}{|} & + & H-O-H & = & \underset{\textstyle O-H}{|} & + & O=N-OH \end{array}$$

Nitrogen peroxide.   Water.   Nitric acid.   Nitrous acid.

These formulæ, which are called constitutional or graphic formulæ, do not in any manner represent the positions in which the atoms are arranged; they are intended to show what atoms are in relations with other atoms in the molecule. We must believe that the atoms in a molecule are in continual motion, which we may compare to the motions of the planets around the sun, and those of the moons around each particular planet. The nature of the molecule depends on all of its atoms, just as the nature of a system of planets depends on the central sun and all the planets and their satellites; and just as the moon would go with the earth were that planet to be withdrawn from the solar system, so do certain groups of atoms enter into the composition of molecules, from which they may separate as groups to form part of other molecules or systems of atoms.

Hereafter we shall not be obliged to use the lines to represent the atomicity of all the atoms in the molecules of which we study the structure. We know that the group hydroxyl, OH, is monatomic, as are also the groups NO and NO^2; on the other hand, we know that the group SO^2 is diatomic, and we can represent our idea that each of these groups exists in the molecule as a distinct part of the system by separating it from the rest of the molecule by a period. Thus we may represent nitric acid by the formula $NO^2.OH$: sulphuric acid, by the formula $SO^2.(OH)^2$.

LESSON XX.

NITRIC ACID. HNO³.

157. Minute quantities of nitric acid often exist in the atmosphere, where they are probably formed under the influence of atmospheric electricity on the nitrogen, oxygen, and moisture of the air. Wherever organized matters containing nitrogen decompose in the presence of porous substances and alkalies, such as potassium hydrate, sodium hydrate, or lime, nitrates are formed. The nitric acid and its compounds of commerce are manufactured from nitrates which are found abundantly in some soils, particularly in India, Egypt, and Chili: in the latter country are large deposits of sodium nitrate.

158. We may prepare some nitric acid by distilling in a glass

FIG. 60.

retort a mixture of sodium nitrate and sulphuric acid, and condensing the vapor in a flask surrounded by cold water. On the large scale, the operation is conducted in cast-iron retorts (Fig. 60), and the vapor is condensed in a series of large stoneware bottles which are called *bon-bons*. As in the decomposition of sodium chloride (§ 75), one molecule of sulphuric acid may be made to decompose either one or two molecules of either potassium or sodium nitrate, forming at the same time either a neutral or an acid sulphate, and setting free one or two molecules of nitric acid.

$$H^2SO^4 \quad + \quad NaNO^3 \quad = \quad NaHSO^4 \quad + \quad HNO^3$$

Sodium nitrate. Sodium acid sulphate. Nitric acid.

$$H^2SO^4 \quad + \quad 2NaNO^3 \quad = \quad Na^2SO^4 \quad + \quad 2HNO^3$$

The proportion required by the last reaction is that employed in the arts, as it is more economical. Let us see what that proportion must be: the molecular weight of sulphuric acid is $\frac{H^2 + S + O^4}{2 + 32 + 64} = 98$; that of sodium nitrate is $\frac{Na + N + O^3}{23 + 14 + 48} = 85$. Then 98 parts of sulphuric acid, and 85 of sodium nitrate, if perfectly pure, would yield $\frac{H + N + O^3}{1 + 14 + 48} = 63$ parts of nitric acid.

Properties.—Nitric acid is a colorless liquid, but is partially decomposed by the prolonged action of light, red vapor being formed and communicating a yellow color to the acid in which it dissolves. It is very volatile, and its vapor condenses the moisture in the air, producing white fumes. Its density is 1.5. It freezes at —49°, and boils at 85°; while boiling it is partially decomposed, so that after a time the boiling point rises to 123°, and a more dilute acid distils, having the same strength as that left in the retort. It mixes with water in all proportions, and the liquid becomes warm during the mixture.

159. By a red heat, nitric acid is at once decomposed into water, red vapor, and oxygen, a decomposition exactly similar to that experienced by lead nitrate under the action of heat (§ 152).

$$2HNO^3 \quad = \quad H^2O \quad + \quad 2NO^2 \quad + \quad O$$

In a small crucible, or a thin iron dish, we heat some powdered charcoal until it becomes barely red hot. We now remove it from the fire, and, when the dish has cooled a little, we pour, at arm's length, some strong nitric acid on the still hot charcoal. At once a vivid combustion takes place; the oxygen of the decomposed nitric acid combines with the carbon, and clouds of red vapor are given off.

On the end of a stick about a metre long we tie a test-tube, into which we pour some strong nitric acid, and if our nitric acid is not the strongest, we add to it about half its volume of sulphuric acid, which will strengthen the nitric acid by its affinity for water. Then in another iron dish we carefully warm some good oil of turpentine until it is nearly boiling. Now we warm our nitric acid, and standing at a distance, pour it suddenly into the hot turpentine: at once the nitric acid oxidizes the turpentine, and, unless the latter has previously become thick by too long exposure to the air, it will be inflamed.

These experiments show us that the oxygen atoms have not exhausted their energy in combining with nitrogen. Indeed, we have seen in the conversion of sulphur dioxide into sulphuric acid that the oxygen of nitric acid is more energetic than in free oxygen molecules at ordinary temperatures, for we have to heat oxygen before it will combine with sulphur dioxide. By reason of this energy still. existing in its oxygen atoms, nitric acid is easily reduced; that is, part or all of its oxygen may be readily removed by oxidizable bodies. We have seen how it is reduced by the hydrogen of another portion of the acid when copper replaces that hydrogen (§ 149): in this same reaction part of the nitric acid is converted into nitrogen monoxide and even free nitrogen, so that the nitric oxide prepared by nitric acid and copper is never perfectly pure. When the reduction by some metals is carried out to its full limit, the nitrogen combines with the hydrogen, forming ammonia. This occurs in the action of zinc on very dilute nitric acid: although zinc nitrate is then formed, no hydrogen is set free, for the displaced hydrogen reduces the nitric acid and combines with the nitrogen: the ammonia formed at once combines with some of the nitric acid present, forming ammonium nitrate.

$$HNO^3 + 4H^2 = 3H^2O + NH^3$$

160. When nitric and hydrochloric acids are mixed, a liquid called *nitro-hydrochloric acid*, or *aqua regia*, is obtained. This liquid is capable of dissolving gold and platinum, a power pos-

sessed by neither of the separate acids. We put a small piece of gold-leaf in a test-tube with nitric acid, and a similar piece in another tube with hydrochloric acid. In neither tube is the gold affected, but on mixing the liquids both pieces are dissolved. Nitro-hydrochloric acid converts the metals into chlorides, the hydrogen of the hydrochloric acid reducing the nitric acid, and the chlorine combining with the metal.

$$2HNO^3 + 2HCl = 2H^2O + 2NO^2 + Cl^2$$

161. Nitrates.—When the hydrogen of nitric acid is replaced by metal, nitrates are formed. We have already learned that an atom of some metals, which we called monatomic metals, is capable of replacing one atom of hydrogen, while the atoms of other metals (diatomic) are able to replace two. Since a molecule of nitric acid contains only one hydrogen atom, an atom of zinc or of lead must replace that atom in two molecules of nitric acid, and consequently it will be united to two groups, NO^3. Then, while we can express the molecules of potassium and sodium nitrates by the formulæ KNO^3 and $NaNO^3$, we must write the molecules of lead and zinc nitrates $Pb(NO^3)^2$ and $Zn(NO^3)^2$.

162. Into a test-tube containing some solution of potassium nitrate in water, we pour a little solution of ferrous sulphate, and then, inclining the tube, some strong sulphuric acid. This last, being much heavier than the other liquids, does not mix at once with the solution, but at the surface, where the sulphuric acid below and the solution of the nitrate touch, a dark ring is formed. This is caused by a partial reduction of the nitric acid by the ferrous sulphate, which produces at the same time a dark color with the nitric oxide resulting from the reduction. This color disappears if we heat the tube (§ 150). This is our test for nitric acid and nitrates.

LESSON XXI.

NITRATES.

163. All of the nitrates are soluble in water. Some of them form anhydrous crystals; others require water of crystallization. When thrown on hot coals, they decompose, and the oxygen given off increases the intensity of the combustion. Salts which so promote combustion are said to *deflagrate* on hot coals.

164. **Sodium Nitrate,** $NaNO^3$, is found in large quantities in Chili and Peru. It forms rhombohedral crystals that are almost cubical; it is very soluble in water. It attracts moisture from the air, and this property prevents its use in the manufacture of gunpowder (see § 166). It is from sodium nitrate that nitric acid and, indirectly, most of the other nitrates are prepared.

165. **Potassium Nitrate,** KNO^3.—This salt is commonly called nitre or saltpetre. In some hot countries it forms an efflorescence, or white powder, on the surface of the soil, and may be obtained by washing the soil with water and evaporating the resulting solution. It is generally made by a double decomposition between the sodium nitrate from Chili and either potassium chloride or potassium carbonate. Boiling solutions of the two substances are mixed, and the sodium chloride or carbonate formed, being much less soluble in boiling water than the potassium nitrate, may be readily separated.

$$KCl \quad + \quad NaNO^3 \quad = \quad NaCl \quad + \quad KNO^3$$
Potassium chloride.   Sodium nitrate.   Sodium chloride.   Potassium nitrate.

$$K^2CO^3 \quad + \quad 2NaNO^3 \quad = \quad Na^2CO^3 \quad + \quad 2KNO^3$$
Potassium carbonate.   Sodium nitrate.   Sodium carbonate.   Potassium nitrate.

Potassium nitrate forms long, six-sided prisms which have a bitter and cooling taste. They dissolve in about five times their weight of water at ordinary temperatures, but require less than half their weight of boiling water. We can now understand how this compound may be separated from sodium chloride, which is about equally soluble in hot and cold water; for when a boiling saturated solution of potassium nitrate is cooled to ordinary temperatures, nine-tenths of the salt separate in crystals, but from a boiling saturated solution of common salt very little is deposited on cooling.

Potassium nitrate deflagrates—that is, increases the activity of combustion—when thrown on hot coals. We melt some zinc in an iron ladle, and, when it is nearly red hot, we throw in a few small pieces of potassium nitrate: the metal takes fire and burns into zinc oxide, the oxygen being supplied from the decomposing potassium nitrate.

166. Gunpowder is a mixture of about seventy-five parts of potassium nitrate, ten of sulphur, and fifteen of charcoal. It is made by grinding each substance separately to the finest powder, and then mixing them and grinding, after a little water has been added. The intimate mixture is then strongly pressed and carefully dried in a warm room, after which it is broken into grains and these are sifted into various sizes. The grains are polished by friction over each other in rotating barrels. This mixture contains all of the materials necessary for its own combustion, and the result of the explosion may be generally expressed by saying that the sulphur combines with the potassium, forming potassium sulphide, while the oxygen of the nitre unites with the carbon to form the gases carbon monoxide and carbon dioxide, which, together with the nitrogen, are set free. The gas occupies a volume very much greater than that of the powder which produced it, and this large volume is made still larger by the high temperature of the reaction. Of course the outside of the grains of powder must burn first, and the larger the grains the slower the combustion and the consequent production of gas; but the smaller the grains the more rapidly is each burned and the flame carried from one to the other. Hence the small quantity of powder used in small-arms is in fine grains in order to produce instantly as much force as possible; but large guns would be broken by such sudden strain, and large grains or lumps are employed, which are put into the gun in coarse bags. In blasting, if it is desired to break the rock in small pieces, a very quickly burning powder is used; but if it is desired to split off

large masses, the effect is accomplished by the more slowly in-
creasing pressure from a slower powder.

• **167. Silver Nitrate,** $AgNO^3$, is made by dissolving silver in
nitric acid, and evaporating the solution until it crystallizes. It
forms colorless plates, soluble in their own weight of water. Its
color darkens by the action of the organic matter in the air and
exposure to light. It melts when cautiously heated, and when cast
into sticks forms the *lunar caustic* used by surgeons. It is a corrosive body, and in the presence of moisture destroys the tissues.
Should any of it by accident be swallowed, common salt is its antidote; insoluble silver chloride is then formed, and this is comparatively harmless (§ 75). When silver nitrate is highly heated,
it leaves a residue of pure silver.

168. Strontium Nitrate, $Sr(NO^3)^2$.—This salt is made by
dissolving the mineral *strontianite*, which is strontium carbonate,
in nitric acid, and purifying by several crystallizations. It forms
colorless crystals, quite soluble in water.

169. Barium Nitrate, $Ba(NO^3)^2$, is obtained like the preceding
salt, but *witherite*—barium carbonate—is used. It also forms colorless crystals, soluble in water, and the solution may be used as a test
for sulphuric acid (§ 115).

170. Cupric Nitrate, $Cu(NO^3)^2 + 3H^2O$, remains in the bottle
in which we prepare nitric oxide. If we filter and evaporate this
solution, the salt separates in large blue prisms; it is very corrosive.
When strongly heated, it leaves black, cupric oxide.

171. Mercuric Nitrate, $Hg(NO^3)^2 + 8H^2O$, separates in large,
colorless crystals when we cool in ice and salt the solution obtained
by boiling mercury in a large quantity of nitric acid. Its solution
is an energetic caustic, and is used in surgery. When dry mercuric nitrate is heated, it decomposes just as the nitrates of lead
and copper, leaving red mercuric oxide.

$$Hg(NO^3)^2 = HgO + 2NO^2 + O$$

172. Lead Nitrate, $Pb(NO^3)^2$.—This compound, which is one
of our most soluble lead salts, is made by boiling lead oxide (litharge) in nitric acid, and evaporating the solution until crystals
separate.

$$PbO + 2HNO^3 = Pb(NO^3)^2 + H^2O$$

It forms colorless, anhydrous crystals, very soluble in boiling water, and in seven times their weight of cold water.

LESSON XXII.

PHOSPHORUS.—HYDROGEN PHOSPHIDE.

173. Phosphorus, P=31.—The element phosphorus is extracted from bones, in which it exists in a compound known as calcium

FIG. 61.

phosphate. The bones are first burned, to remove all of the animal matters, and, by a process which we will understand better when we study the acids of phosphorus, the calcium phosphate which remains is converted into calcium metaphosphate. This last body is mixed with charcoal and strongly heated in clay retorts, and the phosphorus vapor is condensed in vessels containing cold water (Fig. 61). In the reaction which takes place, only half of the phosphorus is separated from the bone-ash, and there is left in the retorts a compound called calcium pyrophosphate, while the gas carbon monoxide is disengaged. The reaction is somewhat complicated.

$$2Ca(PO^3)^2 \quad + \quad 5C \quad = \quad Ca^2P^2O^7 \quad + \quad 5CO \quad + P^2$$
Calcium metaphosphate.   Carbon.   Calcium pyrophosphate.   Carbon monoxide.

Small particles of charcoal are carried over with the phosphorus, which is purified by enclosing it in chamois-skin bags and melting it under warm water. The melted phosphorus is then squeezed through the leather, and so purified is drawn up into glass tubes, where it is allowed to harden in the form of sticks. It is always kept under water, and is transported in sealed tin cans.

174. PROPERTIES.—Phosphorus is an almost colorless, wax-like solid. It is flexible, and soft enough to be readily scratched by the finger-nail. When it has been exposed to light for a long time, its surface becomes white and opaque; it is covered with little crystals of phosphorus; these become loosened, and if we shake the bottle in the dark the whole of the liquid is luminous. Phosphorus has a peculiar, somewhat garlicky odor. Its density is 1.83. It melts at 44°, and boils at 290°. Its density compared to hydrogen is 62; the densities of its gaseous compounds, and the composition of all its compounds, show that its atomic weight is 31; therefore the molecule of phosphorus vapor must contain four atoms, if the molecule of hydrogen contains two. That is, two volumes of hydrogen weighing 2 represent two atoms, but two volumes of phosphorus vapor weighing 124 contain four atoms. A number of other elements also have molecules containing four atoms. Phosphorus is insoluble in water, but dissolves slightly in most oils: it dissolves freely in carbon disulphide, and separates in small crystals when the solution is evaporated very slowly. Phosphorus is luminous in the dark, and this phenomenon is probably caused by a slow oxidation.

175. Phosphorus has an energetic affinity for oxygen. If we expose to the air a small piece of dry phosphorus on a plate, after a time the heat developed by the slow combustion is sufficient to ignite the phosphorus, which takes fire at a temperature of 50°. If we pour on a piece of dry paper on a plate a few cubic centi-

metres of a solution of phosphorus in carbon disulphide, the latter evaporates, leaving the phosphorus in a state of fine division. These small particles are surrounded by oxygen, and the temperature quickly rises till they burst into flame.

Phosphorus is very poisonous: even when poisoning by it is not rapidly followed by death, dangerous diseases of the liver, heart, kidneys, and tongue are produced, and these are usually fatal.

176. **Amorphous Phosphorus.**—The properties which have just been described are those of the common form of phosphorus, but there is another form which may be obtained by heating ordinary phosphorus for a long time to 240°. It then becomes brownish red, opaque, and amorphous. It is not luminous in the dark, does not melt at 44° nor take fire at 50°, is insoluble in carbon disulphide, and is not poisonous. We may easily make a little of this red phosphorus. We put a piece of dry phosphorus in a test-tube, and drop on it a very small flake of iodine: the iodine combines violently with part of the phosphorus, producing light and heat; but the remainder of the phosphorus has become a hard black mass, to extract which we must probably break the tube. This black substance is amorphous phosphorus, and when powdered is brown.

While amorphous phosphorus does not take fire as readily as ordinary phosphorus, its chemical properties are unchanged. We mix a small quantity of *moist* amorphous phosphorus with powdered potassium chlorate, and distribute the mixture on several pieces of paper which we set aside to dry. When quite dry, the least pressure on the spot containing the mixture will cause the oxidation of the phosphorus and decomposition of the potassium chlorate with a loud explosion.

When amorphous phosphorus is heated to 260°, it again changes into ordinary phosphorus.

177. Large quantities of phosphorus are employed for the manufacture of matches. The flame of phosphorus alone would not ignite the stick, because this would become coated with the phosphoric oxide formed, and the latter is a bad conductor of heat. Common matches are therefore first tipped with a paste of

F 11

sulphur, which may take fire from the phosphorus, and the ends of the sticks are then dipped in a paste of ordinary phosphorus with strong glue and some coloring matter. The brown-headed or parlor matches are tipped with a paste made of amorphous phosphorus and potassium chlorate, and sometimes antimony sulphide. The safety matches, which light only on the box, contain the potassium chlorate and antimony sulphide, and these are ignited by friction with amorphous phosphorus glued to the side of the box.

Burns by phosphorus are quite painful and difficult to heal. They are really poisoned wounds, for part of the metaphosphoric acid (§ 187), formed by the action of the phosphoric oxide on the skin, is absorbed, and the gravity of the burn is much greater than that of an ordinary burn of the same size. Phosphorus should always be cut under water, and removed from the water and dried between folds of filter-paper only at the instant before using.

178. Hydrogen Phosphide, PH^3.—In a small glass retort which we have completely filled with a rather strong solution of

sodium hydrate, we put some small pieces of phosphorus, and after arranging the beak of the retort under the surface of water contained in a small vessel, we apply a gentle heat (Fig. 62). When the liquid begins to boil, bubbles of gas rise through the water, and as each bubble comes

FIG. 62.

into the air it takes fire and produces a ring of white smoke. When the air is perfectly still, we notice the curious motions of the rings. The gas which is being formed is hydrogen phosphide, having the composition PH^3, and as it burns the hydrogen is converted into water, and the phosphorus into phosphoric oxide which forms the wreaths of smoke. Hydrogen phosphide is not, however, the only product of the reaction. Part of the phosphorus has been oxidized at the expense of some decomposed water, and the sodium has entered into the new molecule. As we know by analysis that only sodium hypophosphite, having the composition NaH^2PO^2, and hydrogen phosphide are formed, we may write the rather difficult reaction,

$$3NaOH \quad +^- P^4 \; + \; 3H^2O \quad = \quad 3NaH^2PO^2 \quad + \quad PH^3$$

Sodium hydrate. Sodium hypophosphite.

We must notice that the molecule of hydrogen phosphide has a composition like that of ammonia, NH^3. Indeed, it will under proper conditions combine directly with acids, like ammonia, and its compounds, which then contain the group PH^4, are called phosphonium salts.

If we heat red hot in an earthen crucible some fragments of quick-lime, and, having a cover for one crucible, throw in some pieces of phosphorus, covering the crucible after introducing each piece, a calcium phosphide is formed in the crucible. When the crucible and contents have cooled, we may throw some of the pieces of the calcium phosphide into water; bubbles of hydrogen phosphide then come to the surface and take fire spontaneously, forming wreaths of smoke as before. Pure hydrogen phosphide does not take fire on coming into the air, unless the water through which it passes is boiling. That which we have just prepared contains a trace of another compound of phosphorus and hydrogen which is spontaneously inflammable.

179. **Phosphorus Chlorides.**—There are two chlorides of phosphorus. Phosphorus trichloride, PCl^3, is a volatile, colorless liquid. It is made by passing chlorine over phosphorus and condensing the vapor which distils. Phosphorus pentachloride, PCl^5, is a pale yellow, crystalline solid. It is obtained by passing chlorine into the trichloride until the whole becomes solid. Both of these bodies are decomposed by water, as we shall presently see.

LESSON XXIII.

OXIDES AND ACIDS OF PHOSPHORUS.

180. There are two oxides of phosphorus, a trioxide, P^2O^3, and a pentoxide, P^2O^5. The trioxide is formed when phosphorus is slowly oxidized in dry air. The pentoxide, often called phosphoric oxide, results when phosphorus is burned in a full supply of air

or oxygen. We place a piece of phosphorus in a small dish on a plate, and, after igniting it, cover the dish with a bell-jar. In a short time the phosphoric oxide formed settles on the dish and sides of the jar in the form of a snowy-white powder. When we sprinkle some drops of water on this powder, a hissing noise is heard; the water and phosphoric oxide combine, producing much heat and an acid of phosphorus. The composition of the acid which is formed depends on the quantity of water present, for one molecule of this same phosphoric oxide is able to react with one, two, or three molecules of water, forming three different acids.

$$P^2O^5 + H^2O = 2HPO^3, \quad \text{Metaphosphoric acid.}$$
$$P^2O^5 + 2H^2O = H^4P^2O^7, \quad \text{Pyrophosphoric acid.}$$
$$P^2O^5 + 3H^2O = 2H^3PO^4, \quad \text{Orthophosphoric acid.}$$

We shall presently study these acids. Besides these there are two others. Of one we have seen the formation of a salt, sodium hypophosphite; the corresponding acid is of course hypophosphorous acid, H^3PO^2. The other is formed by the reaction of phosphorus trioxide on water, and one molecule of the trioxide reacts with three molecules of water, forming two molecules of phosphorous acid.

We then have a series of acids.

H^3PO^2, Hypophosphorous acid.
H^3PO^3, Phosphorous acid.
H^3PO^4, Orthophosphoric acid.
$H^4P^2O^7$, Pyrophosphoric acid.
HPO^3, Metaphosphoric acid.

181. Hypophosphorous Acid may be made by boiling phosphorus with barium hydrate, $Ba(OH)^2$, and by the cautious addition of sulphuric acid exactly precipitating the barium from the barium hypophosphite formed. After filtering, the liquid is concentrated until a thick syrup is obtained. This is hypophosphorous acid. Although a molecule of this acid contains three atoms of hydrogen, only one of those atoms is replaceable by metal. It is a monobasic acid, and its salts with a monatomic metal like sodium will contain one atom of metal and the group H^2PO^2. The hypophosphites of diatomic metals must contain two of these groups in order that two atoms of hydrogen may be replaced: barium hypophosphite will, then, be $Ba(H^2PO^2)^2$.

182. Phosphorous Acid is most quickly prepared by the re-action of phosphorus trichloride with water, one molecule of the trichloride requiring three molecules of water.

$$PCl^3 + 3H^2O = H^3PO^3 + 3HCl$$

It is a dibasic acid: it contains two atoms of replaceable hydrogen; we may have a sodium phosphite, Na^2HPO^3, and a sodium acid phosphite, NaH^2PO^3. Barium phosphite would be $BaHPO^3$.

Both hypophosphorous and phosphorous acids have reducing properties; that is, they will take away oxygen from oxidized bodies, so becoming converted into phosphoric acid. Into a test-tube containing a solution of silver nitrate, we pour some solution of sodium hypophosphite: in a short time the interior of the tube is coated with metallic silver by the reducing action of the hypophosphite.

183. Orthophosphoric Acid.—We have seen how this acid, which is commonly called *phosphoric acid*, may result from the action of water on phosphoric oxide. It is also formed by the reaction of phosphorus pentachloride with water.

$$PCl^5 + 4H^2O = H^3PO^4 + 5HCl$$

It is usually made by boiling amorphous phosphorus with nitric acid, which is reduced, red vapors being given off. The liquid is then evaporated to a small bulk, and put in a bell-jar over a dish containing sulphuric acid, which gradually absorbs the remaining moisture. In this manner hard, transparent, and deliquescent crystals of orthophosphoric acid are obtained.

Orthophosphoric acid is tribasic: its molecule contains three atoms of replaceable hydrogen. Consequently it may with the same metal form three different salts, accordingly as one, two, or three atoms of hydrogen are replaced by a corresponding quantity of the metal. The names of these salts should indicate the number of hydrogen atoms which have been replaced, or the number of metallic atoms which have replaced the hydrogen: thus, since one atom of sodium always replaces one of hydrogen, monosodium phosphate is NaH^2PO^4, disodium phosphate is Na^2HPO^4, and tri-

11*

sodium phosphate is Na^3PO^4. We have already learned by several reactions (§§ 118, 136) that one atom of calcium is capable of replacing two atoms of hydrogen; and if we perfectly neutralize orthophosphoric acid with lime (calcium oxide), we must have two molecules of the acid and three of lime.

$$2H^3PO^4 + 3CaO = Ca^3(PO^4)^2 + 3H^2O$$

The tricalcium phosphate so formed is the compound existing in bone-ash, from which phosphorus is obtained. It is insoluble in water; when it is treated with sulphuric acid, two atoms of calcium are taken from its molecule, forming calcium sulphate, while calcium acid phosphate passes into the solution.

$$Ca^3(PO^4)^2 + 2H^2SO^4 = CaH^4PO^4 + 2CaSO^4$$
Tricalcium phosphate.   Calcium acid phosphate.   Calcium sulphate.

The calcium sulphate, being insoluble, is separated by filtration, and the calcium acid phosphate is converted into calcium metaphosphate by the action of heat, which decomposes it with the formation of water.

$$CaH^4PO^4 = Ca(PO^3)^2 + 2H^2O$$
Calcium acid phosphate.   Calcium metaphosphate.

184. To a solution of disodium phosphate—either of the other orthophosphates would answer—we add a little ammonia-water, and then some magnesium sulphate solution. A white precipitate forms; this contains both ammonium and magnesium; two atoms of hydrogen in phosphoric acid are here replaced by one atom of magnesium, and the other by the ammonium group, NH^4.

$$Na^2HPO^4 + MgSO^4 + NH^3 = Na^2SO^4 + Mg(NH^4)PO^4$$
Disodium   Magnesium   Sodium   Ammonio-
phosphate.   sulphate.   sulphate.   magnesium phosphate.

In another test-tube we mix some solutions of disodium phosphate and silver nitrate. A yellow precipitate of trisilver phosphate forms.

$$Na^2HPO^4 + 3AgNO^3 = Ag^3PO^4 + 2NaNO^3 + HNO^3$$

These reactions enable us to identify orthophosphoric acid and the orthophosphates.

185. ORTHOPHOSPHATES.—Disodium phosphate exists in the blood, and the phosphorus which is eliminated from our bodies is

principally in monosodium phosphate, which passes out in the urine. The phosphates containing only one atom of metal redden blue litmus, and are generally called acid phosphates. Those containing two atoms of metal do not affect litmus, and are generally called neutral or common phosphates; while those having three atoms of metal turn red litmus to blue.

Large mineral deposits of tricalcium phosphate exist in many localities, and it is probable that they have been formed from accumulations of bones during prehistoric ages. The mineral *apatite*, generally green in color, is principally tricalcium phosphate.

186. Pyrophosphoric Acid.—When orthophosphoric acid is long heated to a temperature of about 213°, it undergoes partial decomposition : two molecules lose one molecule of water, and then combine together, forming a molecule of pyrophosphoric acid.

$$2H^3PO^4 = H^2O + H^4P^2O^7$$

We can understand this better if we consider the structure of the molecule of phosphoric acid: it must contain three hydroxyl groups, and the other atom of oxygen must be combined directly with the phosphorus atom. By the removal of the elements of one molecule of water, two groups, each containing one phosphorus atom, one oxygen atom, and two hydroxyl groups, will be cemented, we may say, by an atom of oxygen.

$$
\begin{array}{ccccc}
OH & OH & & OH & OH \\
HO-P=O + HO-P=O & = & O=P-O-P=O & + HOH \\
OH & OH & & OH & OH
\end{array}
$$

Two molecules orthophosphoric acid.   Pyrophosphoric acid.

We see then that in certain compounds, such as hydrogen phosphide and phosphorus trichloride, phosphorus is triatomic, but that in other cases, and these are the most numerous, it is pentatomic, or equivalent to five atoms of hydrogen.

We mix some solutions of sodium pyrophosphate and silver nitrate; instantly a white precipitate of insoluble silver pyrophosphate is formed.

$$Na^4P^2O^7 + 4AgNO^3 = Ag^4P^2O^7 + 4NaNO^3$$

187. Metaphosphoric Acid.—When either orthophosphoric or pyrophosphoric acid is heated to redness, water is formed, and there remains a hard, glass-like mass of metaphosphoric acid.

$$H^3PO^4 = HPO^3 + H^2O$$

If an acid phosphate is heated in the same manner, it undergoes a similar decomposition, and a metaphosphate remains (§ 183).

Metaphosphoric acid quickly coagulates or renders insoluble the albumen of white of egg, a property which distinguishes it from both ortho- and pyrophosphoric acids.

Metaphosphoric and pyrophosphoric acids and their salts are poisonous, as are also hypophosphorous and phosphorous acids, but orthophosphoric acid and the orthophosphates are not poisonous unless in such concentrated form as to be corrosive.

188. When either metaphosphoric or pyrophosphoric acid, or any of their salts, is boiled with nitric acid, orthophosphoric acid or one of its salts is formed. We have already seen that phosphorus itself is oxidized to orthophosphoric acid by nitric acid. If to this solution in nitric acid we add a solution of ammonium molybdate also in nitric acid, at once or after a time a bright-yellow precipitate of a body called ammonium phosphomolybdate separates. In this manner we can detect the presence of phosphorus or any of its componnds.

LESSON XXIV.

ARSENIC. As = 75.

189. Arsenic is found associated with many metals, copper, silver, bismuth, nickel, but it is obtained principally from one of its minerals, which contains also iron and sulphur. This mineral is called *mispickel*, and its composition may be represented by the formula $FeSAs$. When it is strongly heated, the arsenic is driven out, and iron sulphide, FeS, remains. The operation is conducted in clay retorts, and the arsenic condenses in sheet-iron receivers. This impure arsenic is generally sold under the name *cobalt;* it is purified by being redistilled out of contact with air.

190. In a small test-tube we heat some commercial arsenic, and soon a bright steel-gray ring forms in the cooler part of the tube; after a time the interior of the ring becomes lined with small but

brilliant metallic crystals. This is the appearance of arsenic, but its surface oxidizes after some exposure to the air, and becomes tarnished. The density of arsenic is 5.7. It does not melt when heated ; it sublimes; but it may be melted to a transparent liquid by heating it under pressure. It is insoluble in water, but is slowly oxidized by the air dissolved in the water, and the oxide dissolves, rendering the water poisonous. When arsenic is heated in contact with air, it volatilizes, and its vapor is oxidized to white arsenious oxide. Arsenic takes fire spontaneously in chlorine, burning into arsenic chloride, $AsCl^3$, which is a volatile, very poisonous liquid.

A small quantity of arsenic is added to the lead for making shot ; it hardens the shot, and the interior of the gun-barrel does not become coated with lead by friction with that soft metal.

191. Arsenious Oxide, As^2O^3.—We heat a very small fragment of arsenic in a test-tube, and presently a white ring condenses on the sides of the tube. The arsenic has been oxidized, and the volatile oxide has condensed in the tube : if we examine the ring by the aid of a good microscope, we find that it is composed of small, eight-sided crystals. These are arsenious oxide. Arsenious oxide is manufactured in this manner, by heating arsenic in contact with the air, and the vapor is condensed either in cool chimneys or in large rooms. As it is a very poisonous substance, the operation is conducted with all possible precaution that the workmen may not inhale the vapors and dust.

When it is freshly sublimed in large masses, arsenious oxide is a glassy, transparent, and amorphous solid, but it soon becomes opaque, and this is due to the formation of little crystals. It is not very soluble in water, and the amorphous form is more soluble than the crystalline or opaque variety. Amorphous arsenious oxide dissolves in twenty-five times its weight of cold water, but the crystalline form requires eighty times its weight. The solution contains arsenious acid, but when we evaporate the liquid and try to separate this acid, it is again decomposed into arsenious oxide and water.

$$As^2O^3 \quad + \quad 3H^2O \quad = \quad 2H^3AsO^3$$

Arsenious oxide.   Water.   Arsenious acid.

i

Because arsenious oxide is frequently the cause of intentional or accidental poisoning, it is important that we shall be able to recognize it; but we will better understand its tests when we have learned something of the other compounds of arsenic.

192. Arsenic Oxide and Acids.—When arsenic or arsenious acid is boiled with nitric acid, it is oxidized just as phosphorus was oxidized, and the *ortho-arsenic acid* formed corresponds exactly to orthophosphoric acid. It contains H^3AsO^4. When it is heated to 150°, it is decomposed like orthophosphoric acid, and *pyroarsenic* acid, $H^4As^2O^7$, is formed. This also is decomposed at 200°, yielding *metarsenic* acid, $HAsO^3$, which when heated to redness loses the elements of water, and leaves arsenic oxide, As^2O^5.

$$2H^3AsO^4 \; = \; H^4As^2O^7 \; + \; H^2O \qquad H^4As^2O^7 \; = \; 2HAsO^3 \; + \; H^2O$$
$$2HAsO^3 \; = \; As^2O^5 \; + \; H^2O$$

193. We boil a few grains of arsenious oxide with a few drops of nitric acid in a test-tube, and when the last particle of the solid disappears, we carefully neutralize the liquid with ammonia. Now when we add some silver nitrate solution, a brick-red precipitate of silver arsenate is formed.

$$\underset{\text{Ammonium arsenate.}}{(NH^4)^3AsO^4} \quad + \quad \underset{\text{Silver nitrate.}}{3AgNO^3} \quad = \quad \underset{\text{Silver arsenate.}}{Ag^3AsO^4} \quad + \quad \underset{\text{Ammonium nitrate.}}{3NH^4NO^3}$$

194. Arsenic Sulphides.—In a test-tube of hard glass we melt together some powdered arsenic mixed with a little more than half its weight of sulphur. After cooling, the liquid solidifies to a red mass of *arsenic disulphide*, As^2S^2. This substance is commonly called realgar. It is found as a mineral in transparent red prisms. It is insoluble in water. When heated in the air, both its arsenic and sulphur burn, yielding arsenious oxide and sulphur dioxide.

In another test-tube we melt a mixture of powdered arsenic with about two-thirds its weight of sulphur. When this tube cools, we find in it yellow *arsenic trisulphide*, As^2S^3, generally called orpiment. This sulphide also is found as a mineral. It is insoluble in water, but if boiled for a long time with that liquid it is decomposed, yielding hydrogen sulphide and arsenious acid.

$$As^2S^3 \; + \; 6H^2O \; = \; 2H^3AsO^3 \; + \; 3H^2S$$

Conversely, by passing hydrogen sulphide through a solution of arsenious oxide to which a drop of hydrochloric acid has been added, yellow arsenious sulphide is precipitated.

195. Tests for Arsenic.—Arsenious oxide and some of its compounds are the usual forms in which we must identify arsenic. In a porcelain evaporating dish we heat some pure water, and when it boils we add a few drops of hydrochloric acid, and then put in a thin strip of bright copper. The metal does not tarnish; but when we add to the boiling liquid a little of any solution containing arsenic, the copper soon becomes coated with a steel-gray or even black deposit, which is a compound of copper and arsenic. This is called Reinsch's test. We take this slip of copper from the liquid, wash it in pure water, and carefully dry it between folds of warm filter-paper. Then we cut it into several very narrow slips, and put one or two of these in a little tube drawn out and sealed at one end. We cover this

Fig. 63.

piece of copper with some warm charcoal powder, and then heat the end of the tube. The heat drives the arsenic away from the copper, and the charcoal prevents the vapor from becoming oxidized, so that a gray or black mirror of arsenic condenses in the nearest cool part of the tube (Fig. 63).

196. In another similar tube we put another piece of our coated copper foil, and heat it alone. In this case the arsenic vapor becomes oxidized by the air in the tube, and white arsenious oxide is deposited in minute octahedral crystals that we may recognize when we examine the tube under the microscope (Fig. 64). Were we to break off the portion of the tube containing this deposit, and boil it with a very little water in another tube, we would

obtain a solution of arsenious acid, with which we could make the
next tests; these, however, we will make with larger quantities of
the substance.

197. To a solution of arsenious acid in a test-tube, we add a
drop of ammonia, and then some silver nitrate solution. A canary-
yellow precipitate of insoluble silver arsenite is formed.

$$(NH^4)^2HAsO^3 \quad + \quad 2AgNO^3 \quad = \quad 2NH^4NO^3 \quad + \quad Ag^2HAsO^3$$
Ammonium arsenite.   Silver arsenite.

198. To a similar solution, treated with a little ammonia, we

add cupric sulphate dissolved in water.
An apple-green precipitate of cupric
arsenite, $CuHAsO^3$, is thrown down.

199. In another tube we acidulate
some arsenious acid with a drop of hydrochloric acid, and then pass hydrogen
sulphide through the liquid. A bright
yellow precipitate of arsenic trisulphide
is formed.

Fig. 64.

200. When arsenious acid is poured
into a bottle in which hydrogen is being generated, the nascent
hydrogen, that is, the free atoms of hydrogen which have not
exhausted part of their energy by combining to form molecules,
will reduce or take away oxygen from the arsenious acid, and
combine with the arsenic, forming an exceedingly poisonous gas,
of which we must be careful not to inhale the least quantity. It
is called hydrogen arsenide. Its molecule contains AsH^3.

$$H^3AsO^3 \quad + \quad 6H \quad = \quad 3H^2O \quad + \quad AsH^3$$

We have prepared a hydrogen-bottle with a long jet (Fig. 65),
and, while the hydrogen is burning with its pale flame at this jet,
we pour through the funnel-tube a few drops of a solution of
arsenious oxide. In a few moments the flame becomes bluish and
elongated. Hydrogen arsenide is burning, and the arsenic oxidizes
to arsenious oxide, producing a white smoke. In this flame, and
close to the jet, we hold a plate or piece of cold porcelain, which
will prevent the arsenic from getting enough oxygen to become
oxidized. We see a dark spot of arsenic forming, and we make

several of these spots on different portions of the plate. This is called Marsh's test. If with a lamp we heat the tube of the long jet, the hydrogen arsenide will be decomposed by the heat, and the dark ring of arsenic-deposited in the cooler part of the tube may be afterwards tested as we have already studied.

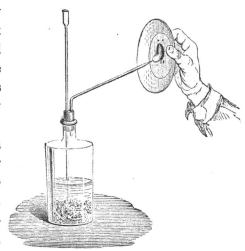

201. We connect a little bent tube with our jet, and pass the gas into some silver nitrate solution in a test-tube; a black deposit of silver separates, and arsenious acid is formed in the solution.

$$AsH^3 + 6AgNO^3 + 3H^2O = H^3AsO^3 + 6HNO^3 + Ag^6$$

We filter the liquid from the silver, and add a drop of ammonia; if all of the silver nitrate has not been decomposed, a yellow precipitate of silver arsenite is formed. We may be obliged to add a few more drops of silver nitrate (§ 197).

202. Now we touch one of the spots on our plate with a drop of strong nitric acid. The spot disappears: we add a small drop of ammonia, and cautiously warm the plate until it is dry. Then we touch it with a drop of silver nitrate, and it becomes brick-red from the formation of silver arsenate (§ 193).

All of these tests enable us to recognize arsenic with certainty; they are applied to substances which are extracted from the body in cases of supposed poisoning.

The green coloring matters known as Scheele's green and Paris green are compounds containing arsenic and copper: they are exceedingly poisonous.

LESSON XXV.

ANTIMONY.   Sb (Stibium) = 120.

203. Antimony is found principally in combination with sulphur in a grayish-black mineral, antimony sulphide, Sb^2S^3. This sulphide is quite fusible, and it is separated from the earthy matters with which it is mixed, and which are called the *gangue*, simply by heating the ore; the antimony sulphide melts and runs out. The easiest method of obtaining antimony from this sulphide is to mix the powdered sulphide with scrap iron and heat the mixture to redness in a crucible. Iron sulphide and antimony are formed, and the latter, being the heavier, collects at the bottom, where we find it as a bright button-shaped lump when we break the cold crucible. The cheapest method, however, is to roast the powdered sulphide; that is, heat it in the air; most of the sulphur is then oxidized to sulphur dioxide, which passes off, and most of the antimony is converted into antimonous oxide. The roasted mass is then mixed with charcoal, and the mixture moistened with sodium hydrate, after which it is heated in crucibles. The carbon removes the oxygen from the antimony oxide, and the sodium hydrate removes the sulphur from the antimony sulphide still present. The sodium sulphide produced forms a slag which floats on the surface of the melted antimony.

PROPERTIES.—Antimony is a very brilliant, white substance, having a high metallic lustre. It is very brittle, and breaks in shining layers: it is said to have a laminated structure. Its density is 6.7. It melts at 450°; when a considerable quantity of it is melted in a crucible and allowed to cool quietly until a crust forms on the surface, if we make a hole in this crust and pour out the still molten interior, the crucible will be found to be lined with small shining crystals.

When antimony is heated in the air, it is oxidized to antimonous oxide, Sb^2O^3. We have already seen that antimony burns spontaneously when thrown into chlorine. It combines with the chlorine, forming antimony pentachloride, $SbCl^5$.

Antimony enters into the composition of several alloys. Type-metal contains twenty per cent. of antimony and eighty per cent. of lead. Lead is too soft for type, and it does not take sharp impressions of moulds: the antimony renders the metal hard, and causes it to expand on solidifying, so filling every line of the mould. Britannia metal also contains antimony.

204. Antimony Chlorides.—By distilling antimony trisulphide with hydrochloric acid, and collecting apart the product which passes after the condensed liquid begins to crystallize in the neck of the retort, antimony trichloride, $SbCl^3$, is obtained as a transparent, colorless solid, melting at 73°, and boiling at 230°. It is soluble in dilute hydrochloric acid, but when the solution is diluted with water, an insoluble oxychloride, $SbOCl$, is thrown down, while hydrochloric acid is formed. Antimony pentachloride, $SbCl^5$, is a volatile, yellow liquid, formed by the action of an excess of chlorine on antimony or the trichloride.

205. Antimony Oxides.—*Antimonous oxide*, Sb^2O^3, is made by heating antimony to redness in open crucibles; after cooling, the latter are found lined with shining, needle-like crystals of the oxide, which corresponds in composition to arsenious oxide. When antimony is boiled with strong nitric acid, it is converted into *metantimonic acid*, $HSbO^3$; by the action of a red heat this is decomposed, yielding *antimony pentoxide*, Sb^2O^5. There is also a *pyrantimonic acid*, $H^4Sb^2O^7$, but there is no orthantimonic acid of the composition H^3SbO^4.

However, this acid, which would have three atoms of replaceable hydrogen, is represented by a curious compound in which we may consider that those three atoms are replaced by a single atom of antimony. When antimonous oxide is heated for a long time in the air, it absorbs oxygen, and the resulting compound has the composition Sb^2O^4, which we may write $Sb(SbO^4)$.

206. Antimony Trisulphide, Sb^2S^3, is the grayish-black mineral from which we have already learned that antimony is obtained. It is a heavy, crystalline substance, having a marked metallic appearance. It is commonly called stibium. This same sulphide may be obtained in another form. Through a solution of antimony trichloride we pass hydrogen sulphide; an amorphous, orange-colored precipitate is formed, and this is antimony trisulphide.

$$2SbCl^3 \quad + \quad 3H^2S \quad = \quad Sb^2S^3 \quad + \quad 6HCl$$
Antimony trichloride.   Antimony trisulphide.

207. When a solution containing antimony is introduced into a bottle in which hydrogen is being generated, some of the hydrogen combines with the antimony, producing a gas, hydrogen anti-monide.   Although this gas has not been obtained in a pure state, being very easily decomposed, enough has been learned about it to show that it has the composition SbH^3.   It causes the hydro-gen to burn with a bluish flame, somewhat like that of hydrogen arsenide, and it also produces dark spots on a piece of porcelain held in the flame, as well as rings in the heated tube (§ 200) ; but here the resemblance with arsenic ceases.   When the spots are oxidized by nitric acid and then treated with silver nitrate, no brick-red color is produced.   When the gas is passed through silver nitrate solution, a dark compound of silver and antimony is precipitated, and, as the clear liquid then contains only nitric acid; it cannot give a precipitate when neutralized with ammonia (§§ 200–202).

208. When we compare the compounds of nitrogen, phosphorus, arsenic, and antimony, we find that the atoms of these elements are almost alike as far as their power of combining is concerned.   One atom of each will combine with three atoms of hydrogen, and we then have formed the four gases,

$$NH^3 \qquad PH^3 \qquad AsH^3 \qquad SbH^3$$

Their more important compounds with chlorine show the same similarity :

$$NCl^3 \qquad PCl^3 \qquad AsCl^3 \qquad SbCl^3$$

But, in addition to these chlorides, phosphorus and antimony form penta-chlorides, PCl^5 and $SbCl^5$.   Each of the four elements has a trioxide and a pentoxide, and from each of the pentoxides is derived an acid containing one atom of hydrogen, one atom of the element, and three atoms of oxygen :

$$HNO^3 \qquad HPO^3 \qquad HAsO^3 \qquad HSbO^3$$

In addition, phosphorus and arsenic form the ortho-acids, H^3PO^4 and H^3AsO^4, while phosphorus, arsenic, and antimony form the pyro-acids, $H^4P^2O^7$, $H^4As^2O^7$, and $H^4Sb^2O^7$.

These similarities and many others enable us to group together the four ele-ments in a natural class ; since in their compounds one atom of either of the class has a combining power equal to that of three or of five atoms of hydro-gen, we may call the class the group of triatomic or pentatomic non-metals.

209. There are three other elements which would be placed in the class that we have just considered, but they occur in such small quantities, although widely distributed, that we can only mention their names.   They are *vana-dium, niobium,* and *tantalum.*   They are found in the minerals vanadanite, columbite, and some others.

LESSON XXVI.

BORON. $B = 11$.

210. The well-known substance, borax, is a compound of the element boron. To a saturated solution of borax we add some sulphuric acid: soon there separates a deposit composed of small white flakes. If we filter these flakes from the liquid, and dry them, we have pearly white scales which feel greasy like soap when we take them between the fingers. This substance is boric acid; it contains H^3BO^3. If we heat it red hot in a platinum crucible, it decomposes and leaves boric oxide, B^2O^3.

$$2H^3BO^3 = 3H^2O + B^2O^3$$

When this boric oxide is mixed with pieces of sodium, some common salt being added to make the mixture melt more readily, and heated to bright redness in a covered iron crucible, sodium borate and boron are formed.

$$3Na^2 \quad + \quad 2B^2O^3 \quad = \quad 2Na^3BO^3 \quad + \quad B^2$$
$$\text{Boric oxide.} \qquad \text{Sodium borate.}$$

After the crucible has cooled, the fused mass is treated with dilute hydrochloric acid, which dissolves the sodium borate, leaving the boron as a dark-brown or olive powder.

Boron is amorphous, infusible, insoluble in water. When it is heated in the air or in oxygen, it takes fire and burns to boric oxide. It is one of the few elements which combine directly with nitrogen: at a red heat in an atmosphere of nitrogen it is converted into boron nitride, BN. It also burns in nitrogen dioxide when heated in that gas, and forms a mixture of boron nitride and boric oxide.

211. **Boric Oxide.**—Boric oxide, of which we have already learned the manner of formation, is a hard, transparent, glass-like substance. It melts at a red heat, and when melted has the property of dissolving many metallic oxides, which communicate various colors to the cooled oxide. We heat to redness the end of a

small platinum wire, and, when it is very hot, we dip it into some boric acid or powdered boric oxide; on again heating this in the flame, it melts to a sort of glass bead, which is perfectly transparent and colorless when cold. We now dip it into a solution of cobalt chloride, and again heat it : when it cools, the bead has a blue color. This blue color is given by the cobalt. As many metals give peculiar colors to such beads of boric oxide, we have in that substance a valuable reagent to aid in the detection of the metals.

Boric oxide is not reduced by heating it with charcoal, but when chlorine is passed over a red-hot mixture of boric oxide and charcoal, carbon monoxide, CO, is disengaged, together with the vapor of a very volatile liquid, boron chloride, BCl^3.

$$B^2O^3 \quad + \quad 3C \quad + \quad 3Cl^2 \quad = \quad 2BCl^3 \quad + \quad 3CO$$
Boric oxide.   Boron chloride.

212. When boric oxide is melted with the metal aluminium, a part of the metal is oxidized, and another part combines with the boron from which the oxygen was removed; there is so formed a complex compound of boron and aluminium, which separates in small crystals when the mass cools. As these crystals are mixed with the excess of solid aluminium, we must remove that metal by boiling in dilute hydrochloric acid. Small octahedral crystals remain undissolved : they were long regarded as crystallized boron. Their composition is not always the same. Their color is yellow, red, or black : their density is about 2.6, and they are almost as hard as diamond : they will scratch rubies, and have sometimes been employed for polishing precious stones.

213. **Boric Acid and Borates.**—We have already seen that boric acid may be formed by the action of sulphuric acid on borax. It dissolves in about twenty-five times its weight of cold water, and the solution is not very strongly acid; it changes blue litmus to a wine color.

It is found in nature in the craters of some volcanoes. In numerous localities in Tuscany gases issue from cracks in the earth, and these volcanic gases contain boric acid. To obtain this body the gas is caused to bubble through the water of little lakes, and when the water is evaporated, the boric acid is left.

Boric acid is tribasic: when it is heated to 100°, it is decomposed into water and *metaboric acid*.

$$H^3BO^3 \quad = \quad HBO^2 \quad + \quad H^2O$$
Boric acid. Metaboric acid.

If the latter be heated to 140° for a time, it is further decomposed into another acid, called *tetraboric*.

$$4HBO^2 \quad = \quad H^2B^4O^7 \quad + \quad H^2O$$
Metaboric acid. Tetraboric acid.

Tetraboric acid is that to which correspond borax and the common borates. In borax, which is *sodium tetraborate*, both of the hydrogen atoms are replaced by sodium.

214. BORAX, $Na^2B^4O^7$, was for a long time obtained principally from Asia and from the boric acid of Tuscany, but within recent years it has been found in large quantities in certain lakes (Borax Lake, Lake Clear) in California. The lakes are dredged, and the mud from the bottom is boiled with water; the borax separates in crystals when the solution is evaporated. When a very concentrated solution of borax cools, it deposits, between 79° and 56°, octahedral crystals in which one molecule of borax is combined with five molecules of water of crystallization; but below 56° it deposits prismatic crystals containing ten molecules of water. As a given quantity of borax will yield a much heavier weight of the latter crystals than of the former, the prismatic crystals are those prepared for commerce. Prismatic borax dissolves in twelve times its weight of cold, or twice its weight of boiling water.

When borax is heated, it loses its water of crystallization more quickly than that water can evaporate; the borax is consequently dissolved in the separated water: it is said to melt in its water of crystallization. As the water is driven off by the heat, it causes the borax to swell up, until it becomes a dry, white, and very light mass. When this is still further heated, it melts to a sort of glass, which possesses the same property of dissolving metallic oxides that we noticed in fused boric oxide. For this reason borax is often used in analysis instead of boric oxide. Because borax dissolves metallic oxides, it is useful in brazing and welding. The surfaces of metal to be welded together become oxidized at the high temperature necessary, and the oxide would prevent their union : a little borax sprinkled on the hot surfaces dissolves the oxide, and the liquid is squeezed out when they are pressed together, leaving clean surfaces which readily unite.

215. We dissolve in alcohol a little boric oxide, or some borax to which a few drops of sulphuric acid have been added. On lighting this alcohol, it burns with a green flame. This test helps us to recognize boric acid or a borate.

LESSON XXVII.

SILICON. Si = 28.

216. Silicon is one of the most abundant elements. In the form of oxide it exists in silica, or quartz, and it forms part of nearly all rocks and minerals. It is not easily obtained in the free state, but it can be prepared in two forms, as an amorphous brown powder and as dark-gray octahedra. It has a strong affinity for oxygen, and is always found combined with that element in nature.

Fig. 66.

217. **Silicic Oxide,** SiO^2.—This compound, generally called silica, is found in many forms. Crystallized, it constitutes the various kinds of quartz, such as rock crystal and amethyst ; amorphous, it forms agate, flint, carnelian, chalcedony. Sandstones and sand are also silica. When it is not colorless, the colors are due to small quantities of metallic oxides. Bock-crystal is pure silicic oxide. It forms hexagonal prisms terminated by six-sided pyramids, and the angles are often curiously modified (Fig. 66). The density of silica is about 2.6.

It is insoluble in water; infusible except in the flame of the oxyhydrogen blow-pipe, and then only imperfectly. It is not deoxidized by reducing agents, such as hydrogen and carbon, even at the highest temperatures. It is scarcely affected by any chemical agents at ordinary temperatures, with the exception of hydrofluoric acid (§ -93). When it is strongly heated with alkaline hydrates or carbonates, it enters into combination with the metal, forming silicates, and these silicates are capable of dissolving silica at very high temperatures. When the mass cools, it constitutes glass, and the properties of the glass depend upon the proportions of silica and alkaline hydrate or carbonate employed. If there be a large proportion of the alkali, the glass is soluble in water, and soluble glass is made by fusing silica with either potassium or sodium carbonate,—generally the latter, because it is cheaper. The solution of this substance hardens as the water evaporates, and is employed as a cement and in making artificial stone.

218. Ordinary glass is made by melting in large clay pots, or in furnaces of peculiar construction, a mixture of fine white sand, sodium carbonate, and lime. If it is desired that the glass shall not soften at a high temperature, potassium carbonate is used instead of sodium carbonate. When the bubbles of carbon dioxide, which are given off from the carbonate employed, have escaped from the pasty liquid, the workman takes out some of the molten metal, as it is called, on the end of a long iron tube, through which he blows air into this lump of soft glass; if the lump be in a bottle-mould, the glass takes the form of the mould. Common window-glass is made by blowing large globes which are drawn out into cylinders by their own weight as they hang on the blow-pipe. The cylinders when cold are cut open their whole length, and are then heated in a furnace, and when soft enough are unrolled into sheets.

219. CRYSTAL, the very heavy and perfectly colorless glass from which cut-glass objects are made, contains lead silicate, which is formed by adding red lead to the mixture of alkaline carbonate and sand before fusing it. A little lead is often used in making common glass. The dark-green color of bottle-glass is caused by

the presence of iron in the sand used, and, in general, colored glasses owe the color to the presence of certain metals, as we shall in time learn. Plate-glass is cast on polished metallic tables, and while still soft is rolled out by heavy rollers,. as dough is rolled. It is afterwards ground flat and polished by machinery. Tumblers, goblets, and like objects are made by pressing the soft glass into moulds.

220. By the action of energetic acids the alkaline metal is at once removed from soluble glass. We pour into a saucer some thick solution of sodium silicate (sodium soluble glass), and on the surface of this we carefully pour some hydrochloric acid; on pouring these liquids from a little height into another saucer, as they run out they mix on the edge, and the silica which is separated from the sodium hangs on the saucer in long icicle-like masses.

In a beaker glass we make a rather dilute solution of sodium silicate, and gradually mix it with dilute hydrochloric acid. Here also sodium chloride is formed by the action of the hydrochloric acid, but no silica is precipitated. Where is it? It must be in the solution, and it exists there in the form of soluble silicic acid. It may be separated from the sodium chloride by a process called *dialysis;* the sodium chloride is a crystalline body, but silicic acid is amorphous. When a solution containing a mixture of crystalline and amorphous bodies is put in a *dialyser,*—which is any glass vessel of which the bottom is cut out and replaced by a piece of parchment paper firmly tied on,—and the dialyser is placed in a vessel of water, the crystalline substance passes through the membrane, while the amorphous body remains in the interior. Then when we pour our solution containing silicic acid into a dialyser, and set the dialyser in a vessel of water, after a time we find the silicic acid alone in the water of the inner vessel. This acid probably has the composition $H^4SiO^4 = 2H^2O + SiO^2$. If we set aside for a few days the beaker containing it, the whole liquid is converted into a jelly. The silicic acid has become an insoluble silicic hydrate, $H^2SiO^3 = H^2O + SiO^2$.

221. **Hydrofluosilicic Acid.**—We have seen how hydrofluoric

acid attacks silica (§ 93). We put into a glass flask an intimate mixture of calcium fluoride (fluor-spar) with fine quartz sand and enough sulphuric acid to make a creamy liquid. We have adapted to our flask a cork having a safety-tube which may pass to the bottom dips into some mercury. On this mercury we pour some water, and, as our gas must overcome the pressure of this water and the mercury, we pour a little mercury in the safety-tube (Fig. 67). We now gently heat our flask, and as each bubble of gas passes through the mercury and touches the water, a gelatinous deposit of silicic hydràte is produced; we use the mercury in order that the delivery-tube may not

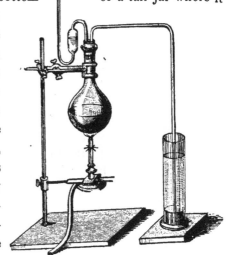

tube and a delivery. of a tall jar where it

FIG. 67.

become stopped by this deposit. In the reaction which is taking place, the hydrofluoric acid which is eliminated from the fluor-spar and sulphuric acid at once acts upon the silica, forming silicon fluoride, $SiFl^4$. This is the gas which comes from the flask.

$$2CaFl^2 + 2H^2SO^4 + SiO^2 = 2CaSO^4 + 2H^2O + SiFl^4$$
Calcium fluoride.   Silicic oxide.   Silicon fluoride.

When this gas comes in contact with water, a reaction takes place, in which silicic hydrate, H^2SiO^3, is formed, together with a gas which dissolves in the water and is called hydrofluosilicic acid. It is a double fluoride of silicon and hydrogen.

$$3SiFl^4 + 3H^2O = H^2SiO^3 + 2(SiFl^4.2HFl)$$
Silicon fluoride.   Hydrofluosilicic acid.

The strong solution of this gas is a highly acid liquid, and is valuable as a test for the metals potassium and sodium, which it precipitates from solutions of their salts. To a solution of potassium nitrate we add some of our filtered liquid, and at once an insoluble double fluoride of potassium and silicon is precipitated, while nitric acid now exists in the solution.

$$2KNO^3 \quad + \quad SiFl^4.2HFl \quad = \quad 2HNO^3 \quad + \quad SiFl^4.2KFl$$

Hydrofluosilicic acid.   Silicopotassium fluoride.

LESSON XXVIII.

CARBON. C = 12.

222. When we compare together a diamond, a piece of charcoal, and a piece of graphite from a lead-pencil, we would not suppose that they have many properties in common ; much less would we think that they are different forms of the same substance. Yet this is the case. They are only varieties of the element carbon. We do not know how the diamond and graphite have been formed in the earth's crust, but we know that many modifications of charcoal may be formed by the decomposition of various compounds of carbon. It is probable that diamonds were formed in the same manner, but under circumstances which we have not yet been able to imitate. We may say, then, that there are three modifications of carbon, two crystalline (diamond and graphite), and one amorphous, under which we must class all the varieties of coal and charcoal.

223. DIAMOND.—This is the hardest of substances : it can be cut and polished only by its own dust. It is found crystallized in regular octahedra and forms of twelve, twenty-four, and forty-eight faces, and the faces are usually curved (Fig. 68). The most highly prized varieties are perfectly colorless, but the tints vary through all the shades, and some diamonds are black. Its density is about 3.5. It is a bad conductor of heat and electricity, and strongly refracts light. When it is strongly heated in a vacuum, it blackens, and is converted into a sort of coke. When strongly heated in oxygen, it burns into carbon dioxide.

224. GRAPHITE, or plumbago, is often called black lead. It

occurs in brilliant black masses, and sometimes in six-sided plates. It is soft enough to be easily scratched by the finger-nail, and leaves a black mark on paper. Its density is 2.2, and it is a good conductor of heat and electricity. It burns into carbon dioxide when heated in air to very high temperatures. It is not usually perfectly pure carbon, but contains one or two per cent. of foreign

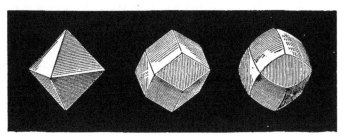

Fig. 68.

matters. Graphite is used in lead-pencils, and for the manufacture of crucibles : in the latter it is powdered and mixed with clay, which binds together the graphite.

225. The other varieties of carbon are amorphous.

ANTHRACITE is a hard and brittle substance, containing from eight to ten per cent. of earthy matters, and sometimes even more.

BITUMINOUS COAL is softer and lighter than anthracite. It contains from 75 to 90 per cent. of carbon, with which is combined a varying proportion of hydrogen. It is the remains of vegetable substances which were buried in the earth in the early geological ages. When it is strongly heated out of contact with the air, various compounds of hydrogen and carbon are formed, together with some water and ammonia. Certain of these compounds of carbon and hydrogen are gases, and, since they contain only combustible elements, they are themselves combustible.

We introduce some fragments of bituminous coal into a small glass retort, to the beak of which we have adapted a little jet (Fig. 69). When we heat the retort by a flame, heavy vapors are disengaged from the coal ; some of them condense in the neck of . the retort, but those which are gaseous at ordinary temperatures

G   k   13

pass out at the jet, and when we apply a flame they burn with a bright light. This is precisely the operation which is conducted

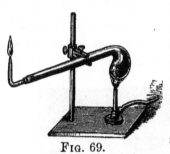

FIG. 69.

for the manufacture of illuminating gas from bituminous coal. The coal is heated in clay or iron retorts (Fig. 70), and the liquid products are condensed by passing through a cold pipe; since coal always contains sulphur and nitrogen, some hydrogen sulphide and ammonia are formed, and these must be separated from the gas, for their combustion would render the air of a room quite unwholesome. They are removed by passing the gas through a tall, upright pipe in

C

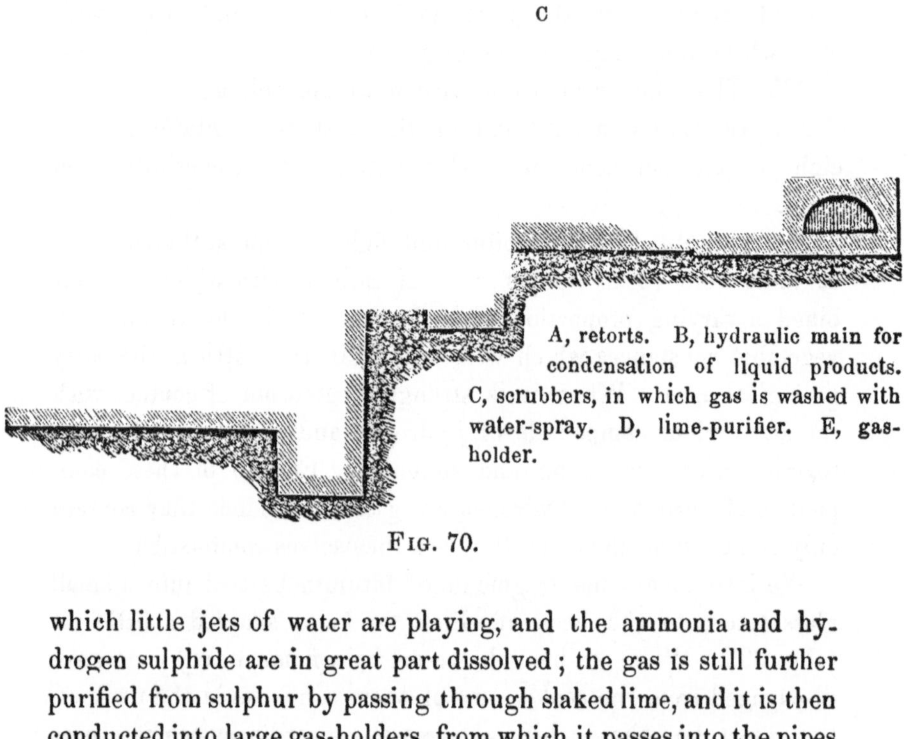

A, retorts. B, hydraulic main for condensation of liquid products. C, scrubbers, in which gas is washed with water-spray. D, lime-purifier. E, gas-holder.

FIG. 70.

which little jets of water are playing, and the ammonia and hydrogen sulphide are in great part dissolved; the gas is still further purified from sulphur by passing through slaked lime, and it is then conducted into large gas-holders, from which it passes into the pipes for consumption. The liquid which first condenses from the gas separates into two layers; one is an impure solution of ammonia, and, together with the water used for washing the gas, forms the source of the ammonia of commerce ; the other is tarry, and contains numerous liquid and solid compounds of carbon and hydrogen, which we must study at another time. The black substance which remains in the clay retorts, as it does in our glass retort, is *coke*. As some of the compounds of hydrogen and carbon which are formed are decomposed by the high temperature of the retorts, these vessels become lined with the carbon, which separates, and forms a dense, hard, strong, and sonorous layer. It is called *gas carbon*, and is used for the carbon plates of voltaic batteries and for the carbon electrodes in the electric light.

The minerals *lignite* and *jet*, of which ornaments are made, are varieties of bituminous coal.

226. CHARCOAL is derived both from wood and from animal matters. Wood-charcoal was formerly made by closely piling the wood and covering the pile with earth, some holes being left for the admission of air. The combustion of part of the wood then produces sufficient heat to convert the rest into charcoal, or carbonize it. This process is very wasteful : not only is a large quantity of wood unnecessarily burned, but the many other products, tar, acetic acid, and wood alcohol, which might be obtained, are lost. Another process is now being everywhere adopted, in which the wood is heated in iron retorts, and the vapors given off are passed through pipes surrounded by other pipes through which flows a stream of cold water; the liquid products are thus condensed, and the gases are conducted under the retort (Fig. 71). The gas from one retort is sufficient to carbonize the wood in another, so that after starting the operation little or no fuel is required and nothing is lost.

Charcoal is brittle and sonorous; its density is about 1.5. It is a poor conductor of heat and electricity. It is not pure carbon : its combustion leaves one or two per cent. of earthy matter, principally the carbonates of potassium and calcium.

ANIMAL CHARCOAL is made by strongly heating waste horn,

bone, blood, hide, and other animal matters, in closed vessels. When made from bone, it is called bone-black or ivory-black: it

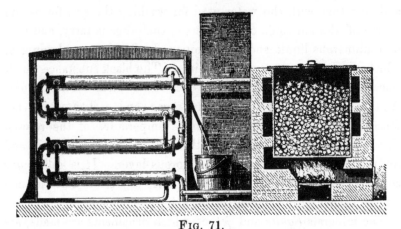

FIG. 71.

then naturally contains the mineral matters of the bones, calcium phosphate and carbonate. These may be dissolved out by washing the bone-black, first with hydrochloric acid, and afterwards with water; it is then called purified animal charcoal.

FIG. 72.

LAMP-BLACK, so much used for the manufacture of printing-ink, india-ink, and black paint, is made by burning oil, turpentine, or rosin in an insufficient supply of air. The operation is conducted in a small furnace of which the chimney opens into a room on whose walls the thick smoke of lamp-black settles. Generally these walls are hung with canvas, from which the lamp-black is removed by a conical scraper which can be lowered by a rope passing over a pulley (Fig. 72). Lamp-black is a fine powder, and usually contains oily and tarry matters from the rosin: it may be purified by heating it red hot in a covered crucible.

227. Properties of Charcoal.—In addition to the peculiarities of each variety which we have considered, all of the forms of charcoal are exceedingly porous, and they are able to absorb many times their volume of certain gases. We fill a rather wide glass

tube with mercury, and, after inverting it in a vessel of mercury, we pass into it some ammonia gas, made by boiling a little ammonia-water in a flask. Now we heat a piece of charcoal red hot, to drive out all of the gases which it has absorbed from the air, and we push it under the mercury into the tube (Fig. 73). It rises to the surface, and instantly we see the volume of ammonia diminishing; the gas is being absorbed by the charcoal, and after we remove the latter from the tube we will find it much heavier, and having a strong odor of ammonia. Charcoal

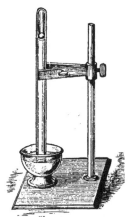

FIG. 73.

will absorb about ninety times its volume of ammonia, fifty-five of hydrogen sulphide, and large quantities of most other gases. It absorbs less than twice its volume of hydrogen, and about eight times its volume of either carbon monoxide, oxygen, or nitrogen. These gases are driven out when the charcoal is heated. We can now understand that in some cases charcoal is an excellent disinfectant: if a dead mouse or other small animal be buried in a box of powdered charcoal, it will be found after some weeks to have dried up, and the charcoal will have absorbed all of the unpleasant odors. Charcoal has also the property of absorbing many coloring matters in its pores, where they are probably oxidized: animal charcoal possesses this property in the most marked degree. We pour some litmus solution into a filter containing animal charcoal, and the liquid passes through colorless. This peculiarity renders

13*

animal charcoal valuable for decolorizing many liquids, and enormous quantities of it are employed for decolorizing sugar. The brown color is removed from crude sugar by dissolving it in water and filtering the syrup through animal charcoal. An excellent filter for the purification of drinking-water consists of a layer of charcoal between two layers of sand : the charcoal must be changed from time to time, or it may be removed and heated red hot in a covered vessel; it is then again fit for use.

228. The strongest affinity of carbon is for oxygen, but this affinity is not manifested at ordinary temperatures. When, however, the temperature is raised to redness and the combustion of charcoal begins, sufficient energy is developed by the chemical action to keep the temperature at the combining point, and the oxidation goes on without further aid. The product of the combustion of carbon in air or in oxygen is carbon dioxide, CO_2. Because of its strong affinity for oxygen, charcoal can remove that element from various oxidized bodies : it is a reducing agent. We have already seen an example of this reduction when we heated charcoal with cupric oxide (§ 13). When such a reduction requires a temperature about redness, carbon dioxide is formed, and this was the case with cupric oxide and charcoal.

$$2CuO \quad + \quad C \quad = \quad Cu_2 \quad + \quad CO_2$$
Cupric oxide.   Copper.   Carbon dioxide.

When, however, the reduction requires a white heat, or near that temperature, carbon monoxide, CO, is formed. This is the action of charcoal on zinc oxide.

$$ZnO \quad + \quad C \quad = \quad CO \quad + \quad Zn$$
Zinc oxide.   Carbon monoxide.

LESSON XXIX.

OXIDES OF CARBON.

229. **Carbon Monoxide,** CO.—This gas might be made by heating to whiteness in clay retorts a mixture of zinc oxide and charcoal, but this would require a furnace and be inconvenient.

We can prepare the gas more conveniently by heating in a glass flask a mixture of oxalic and strong sulphuric acids. The oxalic acid, which is a compound of carbon, oxygen, and hydrogen, is then decomposed into carbon monoxide, carbon dioxide, and water.

$$C^2O^4H^2 \quad = \quad CO^2 \quad + \quad CO \quad + \quad H^2O$$

Oxalic acid.   Carbon dioxide.   Carbon monoxide.

The water will be retained by the sulphuric acid, but we must pass the gases through a bottle containing a solution of sodium hydrate, by which the carbon dioxide will be absorbed. Sodium carbonate is formed in the bottle, and we collect the carbon monoxide over water (Fig. 74).

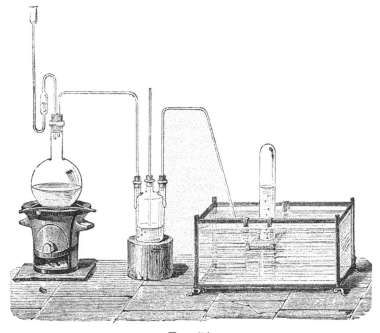

FIG. 74.

230. Carbon monoxide is a colorless, odorless gas. Its density compared to air is 0.967, or compared to hydrogen, 14. It is insoluble in water. It is very poisonous: when it is taken into the lungs it combines with the red globules in the blood, and prevents them from carrying into the system the oxygen which is necessary for the processes of life (§ 33).

Just as we made a similar experiment with hydrogen, carefully we lift our jar from the water, and push into it a lighted taper. The gas takes fire and burns with a blue flame at the mouth of the jar, but the taper is extinguished.   Carbon monoxide will not support combustion, but it will combine with oxygen to form carbon dioxide.

It is interesting here to study the amount of heat disengaged by the combustion of carbon.   Naturally, we can understand that when a given weight of charcoal is burned, a fixed quantity of heat will be developed, enough to raise a certain weight of water, let us say, from 0° to 1°.   When this same weight of charcoal is converted into carbon monoxide, the combustion of the latter gas will not heat as much water through the same temperature.   What has become of the energy which has disappeared from the charcoal?   It must have been lost from the atom of carbon and the first atom of oxygen with which it combined.   Many careful experiments have shown that this is the case, and that the same quantity of carbon always develops the same quantity of heat when it is converted into carbon dioxide, whether it is so converted at once, or whether it first forms carbon monoxide and this combines with an additional atom of oxygen.   We so have a method which enables us to determine the heat or energy of formation of bodies like carbon monoxide. For if from the amount of energy developed by the conversion of a certain amount of carbon into carbon dioxide, we subtract that which is developed by the combustion of a quantity of carbon monoxide containing the same weight of carbon, we will have the energy with which one atom of carbon combines with one atom of oxygen.   In making such determinations, a number of grammes of the substance is taken which would express the atomic weight if one atom of hydrogen weighed one gramme; and the result, which is expressed in the number of kilogrammes of water which would be raised from 0° to 1° by the heat produced, is called the heat of formation of the compound.

231. Carbon monoxide is formed by the action of carbon dioxide on carbon at very high temperatures.

$$CO^2 + C = 2CO$$

When fresh coal is thrown on a hot fire, the escape of the carbon dioxide from the burning coal is retarded, and that gas remains in contact with the coal long enough to be partially reduced to carbon monoxide.   The latter then occasions the blue flame with which we are all familiar.   Carbon monoxide has the property of passing through the pores of red-hot iron, and it often so escapes through the iron of stoves which are not properly lined with fire-brick; fortunately, the gas formed from coal has a decided odor, usually due to the sulphur in the coal, and this odor generally makes us aware of the presence of the poisonous gas.

232. When steam is passed over hot coal or charcoal, a mixture of carbon monoxide and hydrogen is formed.

$$C + H_2O = CO + H_2$$

When this mixture is passed through volatile compounds of hydrogen and carbon, such as we shall learn are contained in the lighter kinds of petroleum, the gases become charged with the vapors of those compounds: if they then be passed through hot pipes, various gaseous compounds of carbon and hydrogen are formed, and these burn with light-giving flames. The water-gas used for illumination in some cities is manufactured in this manner.

233. CARBONYL CHLORIDE.—Carbon monoxide combines directly with chlorine when a mixture of the two gases is exposed to direct sunlight, and a suffocating gas, whose molecule contains $COCl_2$, is formed. The carbon monoxide acts as a radical, just as sulphur dioxide in sulphuryl chloride, SO_2Cl_2, and this radical is called *carbonyl*. Like sulphuryl chloride, carbonyl chloride reacts with water, and in so doing it yields carbon dioxide and hydrochloric acid.

$$COCl_2 + H_2O = CO_2 + 2HCl$$

234. **Carbon Dioxide, CO_2.**—In a gas-bottle, like that which we used for preparing hydrogen (Fig. 75), we put some water and

broken marble, and pour hydrochloric acid through the funnel-tube. As the gas with which we wish to experiment is very heavy, we collect it by downward dry displacement, passing the end of our delivery-tube to the bottom of the jar. When the effervescence in the bottle has continued for a moment, we put a lighted taper into the jar in which we are collecting the gas: the flame is at once

FIG. 75.

extinguished. The jar is full of carbon dioxide, and for such substances as the matter of the taper the oxygen in this gas has exhausted its energy in combining with the carbon : it will unite with no more. In our gas-bottle we have formed water and calcium chloride, for marble is calcium carbonate.

$$CaCO_3 + 2HCl = CaCl_2 + H_2O + CO_2$$
Calcium carbonate.   Calcium chloride.

235. Carbon dioxide is a colorless gas ; it has a faint but somewhat pungent odor and taste. Its density compared to air is 1.529, or compared to hydrogen, 22. We balance on a scale-pan an open and erect paper bag, into which we quickly pour the carbon dioxide from our jar : the descending pan at once shows us that the gas is heavier than the air which it displaces. At 0°, it is converted into a colorless liquid by a pressure of thirty-six atmospheres, and when the liquid is allowed to evaporate rapidly, it absorbs so much heat in assuming the gaseous state that the temperature falls to —78°, and a part of the carbon dioxide is frozen to a snow-like mass. When touched, this solid produces a burn-like fire, for the life of animal tissues cannot continue at such low temperatures.

Carbon dioxide is soluble in its own volume of water, and the quantity of the gas which can be absorbed by a given quantity of water is directly proportional to the pressure : if the pressure be doubled, twice as much gas will be dissolved, etc. We know, however, that by a double pressure two volumes of any gas will be reduced to one volume (Mariotte's law) : hence we may say that water always dissolves its own volume of carbon dioxide, no matter what the pressure. When the pressure is diminished or removed, the gas escapes with effervescence, until the volume remaining dissolved is equal to that of the water. The beverage generally known as gaseous water or soda-water, is simply water into which about five times its volume of carbon dioxide has been pumped.

236. We have already seen that carbon dioxide neither burns nor supports combustion, and a simple experiment shows us its power of extinguishing burning bodies, at the same time that we will be reminded of its weight. We fix a short taper or piece of candle in a cork, and, after lighting it, put it in a small jar into

which we pour the carbon dioxide from another jar which we have filled by dry displacement: the flame is extinguished as if we had poured water on it (Fig. 76). Carbon dioxide is not poisonous, but it produces death by suffocation,—that is, exclusion of oxygen. It collects in wells and brewers' vats, and may then be detected by its power of extinguishing flames lowered into it : if there be enough of the gas present to extinguish or nearly extinguish a flame, it would not be safe for a man to enter such a place before removing the gas, which can be done by agitating the air so that currents may be established.

FIG. 76.

237. Certain substances which have the power of reducing carbon dioxide may burn in the gas. Over a piece of the metal potassium contained in a glass bulb, we pass carbon dioxide that

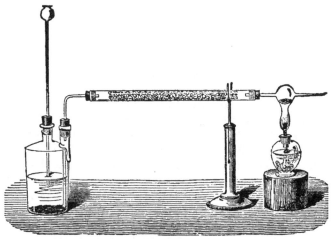

FIG. 77.

has been dried by passing through a tube containing pumice-stone and sulphuric acid. When we warm the potassium, it takes fire and burns with a red light, and a deposit of charcoal is formed in the bulb (Fig. 77). The potassium has reduced some of the carbon dioxide, and has combined with another portion, forming potassium carbonate.

238. We pass a few bubbles of carbon dioxide into some lime-water: the liquid quickly becomes milky by the formation of insoluble calcium carbonate. This test enables us to recognize carbon dioxide. Calcium carbonate is insoluble in water, but if we pass the gas for a long time through our milky liquid, the cloudiness disappears; water containing carbon dioxide in solution will dissolve calcium carbonate. If, however, we boil this liquid so that all of the dissolved carbon dioxide shall be driven out, the calcium carbonate again separates, usually as small crystalline particles, which settle as the water cools. The stalactites and incrustations in caves are formed by the drippings of water holding calcium carbonate in solution by an excess of carbon dioxide; as the gas passes off gradually into the air, the calcium carbonate becomes insoluble and is deposited.

LESSON XXX.

CARBONATES.

239. Carbon dioxide corresponds to an acid which would be formed by the action of one molecule of the gas on one molecule of water. The solution of the gas in water is feebly acid, and we may believe that it contains carbonic acid, $H^2CO^3 = CO^2 + H^2O$. Carbon dioxide is often called carbonic anhydride; anhydride means without water, and carbonic anhydride thus signifies carbonic acid less the elements of water. Although this acid cannot be separated from the solution, for when the water is evaporated carbon dioxide is driven out, there are numerous salts formed by the replacement of one or both of the atoms of hydrogen in H^2CO^3. These salts are the carbonates; they may be easily recognized by the action of hydrochloric acid, which produces with them an effervescence due to the escape of carbon dioxide; we may identify this gas by the milkiness which it produces in lime-water.

There are two classes of carbonates; those in which both atoms of hydrogen in H^2CO^3 are replaced by metal, and others in which only one of these atoms is replaced. The latter are called the acid carbonates, or sometimes the dicarbonates. Carbonic acid is, then, dibasic.

With the exception of the carbonates of sodium, potassium, and lithium, all of the carbonates of the metals are insoluble in water: they dissolve slightly, however, in water containing carbon dioxide. They all effervesce when treated with hydrochloric or sulphuric acid, carbon dioxide being disengaged.

240. **Sodium Carbonate**, Na^2CO^3.—Enormous quantities of this salt, which is commonly called *soda*, or *sal soda*, are used for the manufacture of glass and of soap, and for the preparation of the many compounds of sodium that are used in the arts. It is usually manufactured from common salt, and the process which is coming into general use depends on a reaction between the salt and ammonium acid carbonate (§ 251). We mix saturated solutions of ammonium acid carbonate and sodium chloride, and a fine white deposit is formed in the liquid. This is sodium acid carbonate, and ammonium chloride exists in the solution.

$$NaCl + (NH^4)HCO^3 = NH^4Cl + NaHCO^3$$

| Sodium chloride. | Ammonium acid carbonate. | Ammonium chloride. | Sodium acid carbonate. |

This operation is conducted on a large scale, and the sodium acid carbonate is converted into sodium carbonate by the action of heat. Two molecules of the acid carbonate then lose one molecule of water, and one of carbon dioxide.

$$2NaHCO^3 = Na^2CO^3 + H^2O + CO^2$$

| Sodium acid carbonate. | Sodium carbonate. |

The ammonium chloride is converted into ammonia by heating it with lime (§ 136), and the carbon dioxide formed by heating the sodium acid carbonate, together with more which is obtained from the gases of the furnace-chimneys, serves to convert the ammonia again into ammonium acid carbonate.

$$NH^3 + H^2O + CO^2 = (NH^4)HCO^3$$

The only waste product is, then, the calcium chloride left after the preparation of the ammonia. In the older process, which the ammonia-soda process is gradually replacing, the sodium chloride is first converted into sodium sulphate (§ 77); this is mixed with chalk and coal, and the mixture heated by the flame of a reverberatory furnace (Fig. 78) yields a mixture containing calcium sulphide and sodium carbonate. The last is dissolved out by water,

14

and the waste heat of the furnace is employed not only to evap-orate the solution obtained (C and D), but to dry the mixture of sodium·sulphate, chalk, and coal (B) before introducing it into the hottest part of the furnace (A). This is named, from its inventor, the Le Blanc process.

Sodium carbonate is also manufactured from the mineral cryolite, a double fluoride of sodium and aluminium, of which large quan-

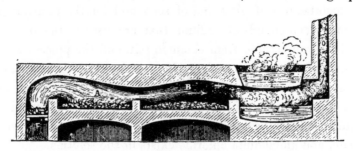

Fig. 78.

tities are found in Greenland. The cryolite is heated with lime, and the reaction yields calcium fluoride and a compound known as aluminate of sodium: it is a combination of the oxides of sodium and aluminium.

$$Al^2Fl^6.6NaFl \quad + \quad 6CaO \quad = \quad 6CaFl^2 \quad + \quad Al^2O^3.3Na^2O$$
Cryolite.   Lime.   Calcium fluoride.   Aluminate of sodium.

The aluminate of sodium is dissolved from the mass by water, and carbon dioxide passed through the solution forms sodium carbonate, while insoluble aluminium hydrate is precipitated.

$$Al^2O^3.3Na^2O \quad + \quad 3CO^2 \quad + \quad 3H^2O \quad = \quad Al^2(OH)^6 \quad + \quad 3Na^2CO^3$$
Aluminate of sodium.   Aluminium hydrate.

The aluminium hydrate is used for the manufacture of alum.

Sodium carbonate forms large crystals containing ten molecules of water of crystallization for one molecule of the salt. When it is exposed to the air, it gradually loses this water, and falls to a dry, white powder: it is said to effloresce. The crystals dissolve in about four times their weight of water at 20°, and the solution has an alkaline reaction and an unpleasant alkaline taste. The salt is insoluble in alcohol.

241. Sodium Acid Carbonate, $NaHCO^3$, is less soluble than the carbonate, and is precipitated when carbon dioxide is passed through a saturated solution of the latter salt. It is usually a white powder, which, when heated or boiled with water, is decomposed into water, sodium carbonate, and carbon dioxide which escapes. It is commonly called bicarbonate of soda, or baking-soda; it forms part of the mixtures known as baking-powders, which contain some acid substance that may react with the acid carbonate, setting free carbon dioxide in bubbles through the dough. Sodium acid phosphate is such a substance; sodium acid carbonate may convert it into either disodium or trisodium phosphate.

$$NaHCO^3 + NaH^2PO^4 = Na^2HPO^4 + H^2O + CO^2$$

242. Potassium Carbonate, K^2CO^3.—This compound is commonly called potash, because it was for a long time derived only from wood-ashes; it is extracted from the ashes by causing water to trickle through them, a process which is called lixiviation. The solution so obtained. is evaporated to dryness, and the residue strongly heated in the air. The potash of commerce contains only from 60 to 80 per cent. of potassium carbonate. The remainder consists of other potassium salts, principally the chloride and sulphate: when these are partially removed by an imperfect purification, the product is called *pearl-ash*.

At Stassfurth, in Prussia, there are large deposits of a double chloride of potassium and magnesium; the mineral is called Stassfurth salt, and contains $KCl.MgCl^2 + 6H^2O$. It is decomposed by boiling with water, and, on cooling, potassium chloride crystallizes from the liquid, while magnesium chloride remains in the solution. Potassium carbonate is now manufactured from this potassium chloride by a method similar to the Le Blanc process for sodium carbonate.

Potassium carbonate is white, and dissolves in less than its own weight of water. It is very alkaline, and has a burning taste. It is deliquescent; that is, it attracts moisture from the air and dissolves in the water so absorbed. It may be obtained in crystals containing two molecules of water by allowing a hot concentrated solution to cool.

243. Potassium Acid Carbonate, $KHCO^3$, is prepared, like sodium carbonate, by passing carbon dioxide through a solution of potassium carbonate. It is less soluble than the latter, and separates from the solution in crystals. Like sodium acid carbonate, it is decomposed by heat, whether it be dry or in solution.

244. Calcium Carbonate, $CaCO^3$.—This substance is one of the most abundant of minerals. As *calcite*, or *Iceland spar*, it forms doubly-refracting rhombohedra: as *aragonite*, it is in right rectangular prisms. It also constitutes marble, limestone, chalk, and the greater part of the matter of shells and corals. When heated to bright redness in open vessels, it is decomposed into carbon dioxide and lime, which is calcium oxide.

245. Strontium Carbonate, $SrCO^3$, constitutes the white mineral *strontianite*.

246. Barium Carbonate, $BaCO^3$, is found crystallized in nature in the mineral *witherite*. The carbonates of calcium, strontium, and barium are precipitated when a solution of sodium or potassium carbonate is added to a neutral solution of any calcium, strontium, or barium salt.

247. Magnesium Carbonate, $MgCO^3$, constitutes the minerals *magnesite* and *giobertite*. *Dolomite* is a double carbonate of calcium and magnesium; it is a magnesian limestone. White magnesia is a variable compound of magnesium carbonate and magnesium hydrate, made by adding an excess of sodium carbonate solution to a boiling solution of magnesium sulphate, and drying the precipitate.

248. Zinc Carbonate, $ZnCO^3$, constitutes the mineral *calamine*, an important ore of zinc.

249. Ferrous Carbonate, $FeCO^3$, is found native in brown crystals as spathic iron.

250. Lead Carbonate, $PbCO^3$, is found crystallized in nature. It is precipitated as an amorphous white powder when any soluble lead salt is treated with sodium carbonate. White lead is a mixture of varying proportions of lead carbonate and lead hydrate. It is manufactured by the joint action of carbon dioxide and vapor of acetic acid on metallic lead. Acetic acid, that is, vinegar, is put into earthen pots, and the lead, either in a rolled sheet or in flat rings, is supported on little projections above the vinegar (Fig. 79). A great number of these pots are prepared, and loosely covered with disks of lead (D): they are then arranged in layers on boards, each layer resting on a bed of refuse bark from tanneries, or of horse-manure. These substances undergo a sort of slow oxidation, and disengage carbon dioxide, which in the presence of the acetic acid converts the surface of the lead into carbonate. We may suppose that lead acetate is first formed, and that this is at once decomposed by the carbon dioxide, forming lead carbonate, while acetic acid is regenerated. When the greater part of the lead has been so changed into carbonate, the pots are opened and the white lead is scraped from the remaining metal.

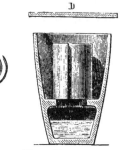

FIG. 79.

There is another process, which depends on the facility with which lead acetate solution dissolves lead oxide, and with which this oxide is precipitated as carbonate when carbon dioxide is passed through the solution. Lead acetate is then again formed in the solution, and is used to dissolve more oxide, which is in its turn precipitated by carbon dioxide.

Excepting sodium carbonate and potassium carbonate, all the other salts which we have just considered are decomposed by heat into carbon dioxide and oxide of the metal.

251. Ammonium Carbonates.—The ammonium carbonate of commerce is commonly called a sesquicarbonate, and is probably a compound of several bodies. It is made by subliming a mixture of chalk and ammonium sulphate in large retorts. Its composition is expressed by the formula $2[(NH^4)^2CO^3] + CO^2 + H^2O.$ It is a crystalline substance, having a strong ammoniacal odor and a sharp, burning taste. It is soluble in water. When it is exposed to the air, it gradually decomposes, losing ammonia and leaving *ammonium acid carbonate*, $NH^4.HCO^3$. The latter body

l

may also be formed by passing carbon dioxide into ammonia-water until no more of the gas is absorbed. It then crystallizes when the liquid is cooled. *Ammonium carbonate*, $(NH^4)^2CO^3$, separates in crystals when the ammonium carbonate of commerce is dissolved in ammonia-water and the solution is artificially cooled.

LESSON XXXI.

COMPOUNDS OF CARBON WITH SULPHUR AND NITROGEN.

252. Carbon Disulphide, CS^2.—When sulphur vapor is passed over red-hot charcoal or coal, the two elements combine, forming carbon disulphide, which passes off as a vapor that may be condensed in any suitable cooling apparatus. This substance is manufactured by throwing sulphur into coal heated to redness in inclined iron cylinders, provided with openings for the introduction of sulphur and the escape of the vapor.

It is a colorless liquid, having the property of highly refracting light. When pure, it has a rather pleasant odor, but the commercial product usually contains small quantities of other compounds which communicate to the liquid a strong and often disgusting odor: it is purified by distillation with lime. Its density at 15° is 1.27. It is almost insoluble in water. It boils at 46°. It is very inflammable: we heat a wire to redness, withdraw it from the flame, and for some time after it has cooled below a visible heat it will still inflame carbon disulphide contained in a small dish. The vapor forms an explosive mixture with the air or with oxygen. The products of the combustion of carbon disulphide are carbon dioxide and sulphur dioxide.

$$CS^2 + 3O^2 = CO^2 + 2SO^2$$

If a few thin iron wires are held in the flame of vapor of carbon disulphide which is heated in a small test-tube provided with a cork and jet, the air and sulphur combine, producing brilliant sparks and molten globules of iron sulphide, which drop from the ends of the wires (Fig. 80).

Carbon disulphide is used for extracting oil from seeds and other matters, for it dissolves fatty substances quickly and in the cold, and may readily be distilled and recovered from the solution, leaving a much larger quantity of oil than could be extracted by pressure. It is used in vulcanizing caoutchouc, an operation which depends on the combination of the caoutchouc with a certain quantity of sulphur, as it is able to dissolve not only the caoutchouc but the sulphur chloride which is used in the operation. We have already seen that carbon disulphide dissolves both sulphur and phosphorus.

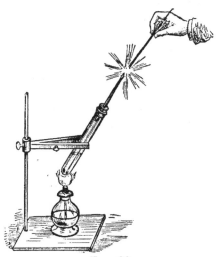

FIG. 80.

253. Carbon disulphide is closely related to carbon dioxide. We have studied the general composition of the carbonates, and know that they correspond to a carbonic acid which should contain H^2CO^3. There is also a series of sulpho-carbonates, exactly similar to the carbonates in composition, but they contain three sulphur atoms instead of three atoms of oxygen.

(H^2CO^3),	Carbonic acid.	H^2CS^3,	Sulphocarbonic acid.
Na^2CO^3,	Sodium carbonate.	Na^2CS^3,	Sodium sulpho-carbonate.
K^2CO^3,	Potassium carbonate.	K^2CS^3,	Potassium sulpho-carbonate.

The sulpho-carbonates have been employed to destroy low forms of life, a purpose for which they are quite effective.

254. If we compare together the molecules of the few compounds of carbon which we have studied, we will find that in all excepting one an atom of carbon has as much combining power as four atoms of hydrogen. It is tetratomic: in carbon dioxide, because it is united with two atoms of oxygen, each of which is worth two atoms of hydrogen; in sodium carbonate, for it is there combined with one atom of oxygen and two other atoms of oxygen, each of which brings into the system a sodium atom as a satellite; in carbon disulphide,

where it is combined with two atoms of diatomic sulphur; and there is also a *carbon oxysulphide*, COS, in which one atom of oxygen and one of diatomic sulphur satisfy the combining capacity of the tetratomic carbon atom. However, in carbon monoxide either the carbon atom must be diatomic,—that is, worth two atoms of hydrogen,—or the oxygen atom must be tetratomic. Since we know that carbon monoxide can combine with two atoms of chlorine, each of which has the power of one hydrogen atom, and that it may also combine with another oxygen atom, we must consider that the carbon atom is diatomic in a molecule of carbon monoxide, which is then an unsaturated compound. We may represent the structure of these molecules, as we have expressed that of others, by structural formulæ :

$$C=O \qquad \underset{\underset{O}{||}}{Cl-C-Cl} \qquad O=C=O \qquad \underset{\underset{O}{||}}{NaO-C-ONa} \qquad S=C=S \qquad O=C=S$$

| Carbon monoxide. | Carbonyl chloride. | Carbon dioxide. | Sodium carbonate. | Carbon disulphide. | Carbon oxysulphide. |

After a time we shall become acquainted with compounds in which the group CS acts as a radical, precisely as does the group carbonyl, CO; so that we may consider the compound COS either as a combination of carbonyl with an atom of sulphur, or as a compound of the group CS with one atom of oxygen (§ 278).

255. In the combining power of its atoms, silicon resembles carbon. It also is tetratomic, as we can understand from the composition of the silicon compounds which we have studied; but there is no monoxide corresponding to carbon monoxide, and silicon does not appear to be diatomic in any compounds.

$$O=Si=O \qquad \underset{\underset{O}{||}}{HO-Si-OH} \qquad \underset{Fl-Fl}{\overset{Fl-Si-Fl}{/\backslash}}$$

| Silicic oxide. | Silicic hydrate. | Silicon fluoride. |

256. Cyanogen, C^2N^2.—

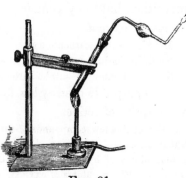

FIG. 81.

We put into a test-tube some mercuric cyanide, a white and very poisonous compound of mercury, carbon, and nitrogen; then we adapt to our test-tube a cork in which we have fitted a bent tube bearing a little bulb containing some small pieces of the metal potassium (Fig. 81); the outer end of this tube is drawn into a fine jet. We now heat the mercuric cyanide, and presently metallic mercury begins to deposit in the cooler part of the tube, and a gas is escaping from the jet: we light it, and it burns with a beautiful peach-blossom-colored flame.

This gas is cyanogen. It is a colorless gas, having an odor like that of bitter almonds, and is quite poisonous. Its density compared to air is 1.8064, or compared to hydrogen, 26; its molecular weight is, then, 52, and analysis has shown that it contains carbon and nitrogen in the proportions indicated for one atom of each. Since its molecular weight is 52, a molecule of cyanogen gas must contain two atoms of carbon and two of nitrogen. Cyanogen is converted into a liquid by pressure or by a temperature of —25°. It dissolves in about one-quarter its volume of water, but the solution soon decomposes, and then always contains ammonia or some ammonium compound. The combustion of cyanogen produces nitrogen and carbon dioxide.

257. By the aid of a spirit-lamp, we now heat the bulb containing the potassium. There is a bright flash of light: the potassium and cyanogen have combined, and formed potassium cyanide. Cyanogen is, then, capable of entering into combination. When, however, we analyze the potassium cyanide formed, we find that it contains potassium, carbon, and nitrogen in the proportions required for one atom of each; its formula is KCN. The cyanogen molecule C^2N^2 has then separated into two groups CN, each of which has combined with an atom of potassium. The group CN is a radical, and free cyanogen resembles free chlorine in this respect, for the molecule of chlorine contains two atoms, while the molecule of cyanogen contains two groups or radicals.

$$\text{Cl-Cl} \qquad \text{NC-CN}$$

The reaction between potassium and cyanogen is, then, like that between potassium and chlorine; both are double decompositions.

$$\text{K-K} + \text{Cl-Cl} \qquad = \text{KCl} + \text{KCl}$$
$$\text{K-K} + (\text{CN})\text{-}(\text{CN}) = \text{KCN} + \text{KCN}$$

258. In free cyanogen gas is it the carbon atoms which are united together, or the nitrogen atoms? When cyanogen or its compounds decompose, the nitrogen atoms always form compounds in which they are triatomic, having the combining power of three atoms of hydrogen. Then we must believe that they

are also triatomic in cyanogen, and since the carbon atom is worth four hydrogen atoms and the nitrogen atom only satisfies three-fourths of this combining power, the carbon atoms must be united together, and we consequently write cyanogen gas $N{\equiv}C{-}C{\equiv}N$ or $(CN)^2$. We see also that the potassium atom in potassium cyanide, and the mercury atom in mercuric cyanide, must be united to the carbon atom of cyanogen. (Compare §§ 262 and 334.)

$$K{-}C{\equiv}N \qquad N{\equiv}C{-}Hg{-}C{\equiv}N$$

Although carbon and nitrogen do not combine directly, potassium cyanide is formed when either nitrogen gas or ammonia is passed over a highly-heated mixture of charcoal with potassium carbonate or potassium hydrate. All of the compounds of cyanogen are prepared from potassium ferrocyanide (§ 266).

LESSON XXXII.

HYDROCYANIC ACID.—CYANIDES.

259. Hydrocyanic Acid, HCN.—This dangerous poison, commonly called prussic acid, is formed when a cyanide is treated with a dilute acid; as by the action of hydrochloric acid on mercuric cyanide.

$$Hg(CN)^2 \quad + \quad 2HCl \quad = \quad HgCl^2 \quad + \quad 2HCN$$
Mercuric cyanide.   Mercuric chloride.

It is usually made by distilling 8 parts of potassium ferrocyanide with a cooled mixture of 9 parts of sulphuric acid and 14 parts of water. The beak of the retort containing this mixture is

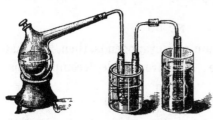

inclined upwards, in order that the water may condense and run back into the retort; the vapor of hydrocyanic acid is dried by passing through a calcium chloride tube placed in water heated to about 30°, and then condensed in a flask surrounded by a mixture of ice and salt (Fig. 82).

Fig. 82.

260. Hydrocyanic acid is a colorless, very volatile liquid; its odor resembles that of bitter almonds. Its density is about 0.7;

it freezes at —15°, and boils at 26.5°. The density of its vapor compared to hydrogen is 13.5, corresponding exactly with the molecular weight implied by the formula HCN. It dissolves in all proportions of water, and a two per cent. solution is used in medicine. It is combustible, and when ignited burns into water, carbon dioxide, and nitrogen. It is exceedingly poisonous, and the accidental inhalation of its vapor has in some cases proved fatal.

261. It is often important to be able to recognize hydrocyanic acid; and we may do so by the following tests. We make our solution of hydrocyanic acid for these tests by adding a little dilute sulphuric acid to some solution of potassium cyanide. The liquid then contains hydrocyanic acid and potassium sulphate.

Over the beaker glass in which we have prepared this solution, we invert a watch-glass or glass plate on which we have placed a drop of silver nitrate solution (Fig. 83): this drop soon becomes clouded from the formation of insoluble silver cyanide; the white deposit does not darken quickly on exposure to light, and is dissolved by nitric acid; these characters distinguish it from silver chloride, which would be formed if the liquid contained hydrochloric acid.

FIG. 83.

We now invert over our beaker another watch-glass containing a drop of ammonium sulphide which has become yellow by exposure to light and air: some ammonia has escaped from it, and it contains an excess of sulphur. In a little while this drop becomes colorless: a compound called ammonium sulphocyanate has been formed in it.

$$(NH^4)^2S \; + \; S^2 \; + \; 2HCN \; = \; 2NH^4CSN \; + \; H^2S$$

Ammonium   Ammonium
sulphide.   sulphocyanate.

If we now carefully warm the spot until it no longer has the odor of hydrogen sulphide, and then touch it with a drop of ferric chloride solution, a blood-red color appears. This color is due to the formation of ferric sulphocyanate (§ 277).

We mix in a test-tube a few drops of our hydrocyanic acid

solution with a little ferrous sulphate and ferric sulphate, and add a little strong solution of sodium hydrate : a dirty deposit forms, but when we add an excess of hydrochloric acid, a part of the deposit is dissolved, and a fine blue precipitate, Prussian blue (§ 267), remains.

262. Hydrocyanic acid does not keep long, soon decomposing, whether it be pure or in solution. It undergoes an interesting reaction with strong hydrochloric acid, and the reaction is more interesting because it is characteristic of all the cyanides. When we mix hydrocyanic acid with strong hydrochloric acid, the mixture becomes hot, and a mass of crystals of ammonium chloride separate. The most curious part of this reaction is, that it takes place between the hydrocyanic acid and the water of the hydrochloric acid; the nitrogen atom of the former is exchanged for an atom of oxygen and a hydroxyl group.

$$HCN \; + \; 2H^2O \; = \; HCO.OH \; + \; NH^3$$

The ammonia formed combines with the hydrochloric acid. The compound HCO.OH is called formic acid. When a solution of potassium cyanide is boiled, it is converted into potassium formate by a similar reaction.

$$KCN \quad + \quad 2H^2O \quad = \quad HCO.OK \quad + \quad NH^3$$

   Potassium cyanide.   Potassium formate.

All the acids of carbon which we shall presently have occasion to study, may be formed by the replacement of the nitrogen atoms of corresponding cyanides by an oxygen atom and a hydroxyl group.

263. Potassium Cyanide, KCN, is made by heating dry potassium ferrocyanide red hot in earthen retorts. After the mass cools, it is extracted with alcohol, and when the filtered liquid is evaporated it leaves a white mass of potassium cyanide. It may be crystallized in cubes. It is soluble in water and alcohol, but the aqueous solution decomposes after a time, even in the cold, into potassium formate and ammonia. When potassium cyanide is heated with sulphur, it is converted into potassium sulphocyanate, CSNK. Solutions of potassium cyanide dissolve the insoluble cyanides of silver, zinc, and other metals, forming double cyanides, which are used in electro-plating. Potassium cyanide is very poisonous, as indeed are nearly all the cyanides.

264. Silver Cyanide, AgCN, is formed as a white precipitate when a solution of silver nitrate is treated with the exact quantity of potassium cyanide required for one molecule of each. When heated, it decomposes into silver and cyanogen gas.

265. **Mercuric Cyanide,** $Hg(CN)^2$.—This compound may be made by dissolving mercuric oxide in dilute hydrocyanic acid, but it is usually prepared by boiling for about fifteen minutes a mixture of one part of potassium ferrocyanide, two parts of mercuric sulphate, and eight parts of water. The mixture is filtered while boiling, and mercuric cyanide separates from the filtrate in colorless, anhydrous, square prisms. It dissolves in eight times its weight of cold water.

266. **Potassium Ferrocyanide,** $K^4Fe(CN)^6$.—Potassium ferrocyanide is the starting-point for the preparation of other compounds of cyanogen. There are a number of processes for its manufacture: the most common of them consists in heating waste animal matters containing nitrogen, such as blood, horn, scraps of skin and leather, with potassium carbonate and scrap iron. After the mass has cooled, it is exhausted with boiling water, and the concentrated solution deposits the ferrocyanide in crystals.

These crystals are yellow, and contain three molecules of water of crystallization, which may be driven out by a temperature of $100°$; the anhydrous salt then remains as a white powder. Crystalline potassium ferrocyanide, which is commonly called yellow prussiate of potash, dissolves in four times its weight of cold or twice its weight of boiling water, and is insoluble in alcohol. It is not poisonous.

The group of atoms $Fe(CN)^6$ which it contains is a radical, and takes part in double decompositions without undergoing change. There is a hydroferrocyanic acid, $H^4Fe(CN)^6$. We add some cupric sulphate to solution of potassium ferrocyanide, and a mabogany-colored precipitate of cupric ferrocyanide is formed, while potassium sulphate goes into solution.

$$2CuSO^4 + K^4Fe(CN)^6 = 2K^2SO^4 + Cu^2Fe(CN)^6$$

Solution of potassium ferrocyanide causes the formation of insoluble ferrocyanides in solutions of many metallic salts, and the color of the precipitate is a means frequently employed for identifying the metals. With zinc sulphate, we would have zinc ferrocyanide, which is white, thrown down.

H   15

When potassium ferrocyanide is heated to redness in closed vessels, it is converted into potassium cyanide, while iron and charcoal separate and nitrogen is disengaged. When it is heated in the air or with certain oxidizing agents, it yields potassium isocyanate, $CONK$. Under the same circumstances with sulphur it forms potassium sulphocyanate, $CSNK$.

267. Prussian Blue, Ferric Ferrocyanide, $(Fe^2)^2(FeC^6N^6)^3$.

—When ferrous sulphate, $FeSO^4$, is added to a solution of potassium ferrocyanide, the atom of iron changes place with two atoms of potassium, and a pale-blue precipitate containing $FeK^2Fe(CN)^6$ is formed. When, however, potassium ferrocyanide is added to a ferric salt, such as ferric chloride, Fe^2Cl^6, a dark-blue precipitate of Prussian blue is thrown down. As the two atoms of iron in ferric chloride replace six atoms of hydrogen in as many molecules of hydrochloric acid, they will also replace six atoms of potassium, and we must write

$$2Fe^2Cl^6 \quad + \quad 3K^4Fe(CN)^6 \quad = \quad 12KCl \quad + \quad Fe^4(FeC^6N^6)^3$$

Ferric chloride. 　　Potassium ferrocyanide. 　　　　　　　　　Prussian blue.

Prussian blue, much used as a pigment, generally comes in cubical masses having a coppery reflection. It is insoluble in water, and in dilute acids, with the exception of solutions of oxalic acid. It is dissolved by alkaline hydrates, which destroy its color.

While we are uncertain of the exact relations of the atoms in the molecules of the ferrocyanides, yet we have learned that they contain a distinct radical, ferrocyanogen, $Fe(CN)^6$; and we see that the relations of the atom of iron in this radical are quite different from those of the four iron atoms in Prussian blue. The latter readily leave and re-enter the molecule by double decomposition, but the iron atom in ferrocyanogen always goes with the six groups, CN, unless the molecule be decomposed by heat or energetic chemical agents. Indeed, two ferrocyanogen groups may combine together, as is the case in

268. Potassium Ferricyanide, $K^6(FeC^6N^6)^2$.

This compound is formed by passing chlorine gas into a solution of potassium ferrocyanide. One atom of potassium is then removed from each molecule of the ferrocyanide, forming potassium chloride, and the remainders of the molecules unite together in pairs, forming potassium ferricyanide.

$$2K^4FeC^6N^6 \quad + \quad Cl^2 \quad = \quad 2KCl \quad + \quad K^6(FeC^6N^6)^2$$

Potassium ferrocyanide. 　　　　　　　　　　　　　Potassium ferricyanide.

Potassium ferricyanide forms beautiful, large, ruby-red, anhydrous crystals. It dissolves in about four times its weight of cold water, and the solution has a greenish-brown color. It forms no precipitate with ferric salts, but with ferrous sulphate gives a dark-blue precipitate of ferrous ferricyanide, called Turnbull's blue.

$$K^6(FeC^6N^6)^2 \ ^- \ + \ 3FeSO^4 \ = \ 3K^2SO^4 \ + \ Fe^3(FeC^6N^6)^2$$

Potassium ferricyanide.   Ferrous sulphate.   Turnbull's blue.

LESSON XXXIII.

CARBONYL COMPOUNDS.

269. We have already seen that carbon monoxide combines directly with chlorine, forming carbonyl chloride, $COCl^2$. To this chloride there corresponds a series of compounds in which the chlorine atoms are replaced by various radicals having the same combining power. The only compounds of this class which we can consider are the isocyanates, urea, and a few closely-allied substances.

270. **Potassium Isocyanate,** CO.NK.—When an intimate mixture of perfectly dry potassium ferrocyanide with half its weight of manganese dioxide is heated to dull redness with constant stirring, the mixture becomes black and pasty. The potassium ferrocyanide has been decomposed, and potassium isocyanate exists in the product. To extract this substance, the black mass is finely powdered and the powder shaken up with boiling eighty per cent. alcohol: the liquid is quickly decanted from the sediment, and on cooling deposits potassium isocyanate in small, colorless, anhydrous crystals. It is very soluble in water; only slightly soluble in cold alcohol. When the aqueous solution is heated, the isocyanate is decomposed into potassium carbonate, carbon dioxide, and ammonia.

$$2CO.NK \ + \ 3H^2O \ = \ K^2CO^3 \ + \ CO^2 \ + \ 2NH^3$$

Potassium isocyanate is decomposed in the same manner by acids: hydrochloric acid converts it into potassium chloride and ammonium chloride, while carbon dioxide escapes with effervescence.

$$CO.NK + 2HCl + H_2O = KCl + NH_4Cl + CO_2$$

The acid corresponding to potassium isocyanate is of course isocyanic acid, $CO.NH$, but it cannot be made by double decomposition with potassium isocyanate.

271. There is another compound whose molecule has exactly the same composition as that of potassium isocyanate, and the corresponding acid, although not yet separated, is cyanic acid. It would contain the radical cyanogen, CN, and a hydroxyl group, OH; the potassium salt representing this molecule, in which the hydrogen atom is replaced by potassium, contains $K-O-C\equiv N$. It has been obtained by the action of a compound called cyanogen chloride, $Cl-C\equiv N$, on potassium hydrate.

$$Cl-C\equiv N + 2KOH = KCl + H_2O + KO-C\equiv N$$

This is the true *potassium cyanate;* the other compound has precisely the same atoms in its molecule, but we shall presently find reasons for believing that these atoms are differently arranged. Compounds whose molecules contain the same number and kind of atoms, and yet have different properties, are called *isomeric compounds.* There are two isomeric potassium compounds whose molecules contain one atom each of potassium, carbon, oxygen, and nitrogen : that containing the radical cyanogen is called potassium cyanite; the other, which we shall see does not contain that radical, is called potassium isocyanate. The relations of the atoms in its molecule are expressed by the formula $O=C=NK$ (§ 274).

272. **Ammonium Isocyanate**, $NH_4.NCO$, is formed when vapor of isocyanic acid is mixed with ammonia gas. It is a white solid, very soluble in water. When its aqueous solution is boiled, or even left to itself for a few days, the ammonium isocyanate is converted into an isomeric body, urea.

273. **Urea**, $CO(NH_2)_2$, may be formed by a reaction which establishes its molecular structure beyond doubt. When carbonyl chloride, $COCl_2$, is made to react with ammonia, urea and hydrochloric acid are formed.

$$COCl_2 + 2NH_3 = CO(NH_2)_2 + 2HCl$$
Carbonyl chloride.   Urea.

Here two molecules of ammonia lose each one atom of hydrogen, which combines with the chlorine of the carbonyl chloride, and the unsatisfied groups CO and $2NH^2$ combine, forming a molecule of urea. The group NH^2 passes readily from one molecule to another by double decomposition. It represents a molecule of ammonia from which an atom of hydrogen has been removed : it is a monatomic radical.‘ We may then consider that urea is formed from two molecules of ammonia by the replacement of one atom of hydrogen of each by the diatomic radical carbonyl. Compounds formed by the replacement of the hydrogen atoms of ammonia by other atoms or groups are called *amines* or *amides :* when the replacement is by the radicals of acids, the name amide is used to designate the new compound, while amine is applied to such compounds as result from the replacement of these hydrogen atoms by radicals which are also capable of replacing the hydrogen of acids. Since carbonyl is the radical of carbonic acid, which is carbonyl dihydrate, $CO(OH)^2$, we call urea *carbonyl amide,* or *carbamide.*

274. We have already learned that urea is formed by a curious change which takes place in ammonium isocyanate. Since we can readily prepare potassium isocyanate, we have a ready means of obtaining urea. For this purpose potassium isocyanate is prepared as has already been described (§ 270); but, instead of exhausting the mass with alcohol, we exhaust it with cold water, which dissolves out the isocyanate. The solution is then mixed with ammonium sulphate in quantity equal to five-sevenths of the potassium ferrocyanide used, and the whole is evaporated to dryness on a water-bath. The ammonium sulphate reacts with the potassium isocyanate, forming potassium sulphate and ammonium isocyanate, and the latter becomes converted into the isomeric compound, urea. The mixture of the two bodies is extracted with a small quantity of boiling alcohol, which does not dissolve the potassium sulphate, but dissolves the urea, and on cooling deposits it in crystals.

Since there can be no question that urea contains the group carbonyl, CO, united to two groups NH^2,—in chemical language, that it is carbonyl amide, or carbamide,—we must infer that ammonium isocyanate, and the potassium isocyanate from which it is derived, also contain the group carbonyl.

$$H^4N–N=CO \qquad\qquad CO(NH^2)^2$$
Ammonium isocyanate.   Urea.

275. Urea is the principal solid constituent of the urine: it is in this compound that the greater part of the nitrogen of burned tissues is removed from the body. It may be extracted from urine by evaporating the liquid to a thick syrup, and adding nitric acid when it has cooled. The nitric acid combines with the urea, forming urea nitrate, $CO(NH^2)^2.HNO^3$, which separates in a mass of crystals. These are drained, and treated with a concentrated solution of potassium carbonate as long as there is effervescence. The mixture is then evaporated to dryness, and the urea is dissolved from the potassium nitrate by boiling alcohol.

276. Urea forms colorless crystals having a cooling taste. It dissolves in its own weight of water, and in five times its weight of cold alcohol; it is very soluble in boiling alcohol. An aqueous solution of chlorine instantly decomposes it, setting free nitrogen and carbon dioxide.

$$CO(NH^2)^2 + H^2O + 3Cl^2 = CO^2 + N^2 + 6HCl$$

By the action of heat, its solution in water is converted into ammonium carbonate.

$$CO(NH^2)^2 + 2H^2O = (NH^4)^2CO^3$$

The same reaction takes place slowly in urine, and accounts for the ammoniacal odor of stale urine.

277. Potassium Sulphocyanate, KN.CS.—A mixture of potassium ferrocyanide with half its weight of flowers of sulphur is heated to dull redness in a covered crucible. After cooling, the mass is dissolved in water, the liquid is filtered, and potassium carbonate is added as long as it causes any precipitate. Then the liquid is again filtered, and the solution evaporated to dryness. The residue is extracted with hot alcohol, and the alcoholic solution allowed to evaporate. Potassium sulphocyanate then separates in colorless, deliquescent crystals which are very soluble in water and in alcohol. A solution of potassium sulphocyanate produces a blood-red color (ferric sulphocyanate) with solutions containing ferric salts.

Potassium sulphocyanate corresponds to the isocyanate in which the oxygen atom is replaced by an atom of sulphur.

278. **Ammonium Sulphocyanate,** (NH⁴)N.CS, is found in small quantity in the water which has been used to wash coal gas (§ 225). Representing ammonium isocyanate in which the oxygen is replaced by sulphur, it undergoes by the action of heat a similar curious change into the isomeric compound *sulpho-urea,* $CS(NH^2)^2$, whose molecule is exactly like that of urea, excepting that it contains sulphur instead of oxygen. It contains the radical CS (§ 254).

LESSON XXXIV.

COMPOUNDS OF CARBON AND HYDROGEN (I).

279. **Methane, CH⁴.**—In a glass flask on a sand-bath we heat a mixture of equal parts of dried sodium acetate, sodium hydrate, and powdered lime (Fig. 84). The lime does not enter into the

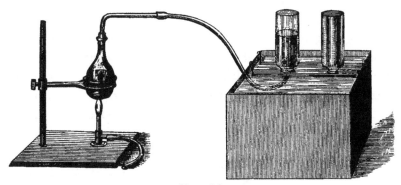

FIG. 84.

reaction which takes place, but it prevents the hot sodium hydrate from melting through the glass. Since gas will be disengaged, we have adapted to our flask a cork and tube, and may collect this gas over water, in which it is almost insoluble. The gas is *methane:* it is produced by a reaction between the sodium acetate and sodium hydrate, which at the same time yield sodium carbonate.

$$NaC^2H^3O^2 \quad + \quad NaOH \quad = \quad Na^2CO^3 \quad + \quad CH^4$$

Sodium acetate.　　　　　　　　　Sodium carbonate.　　Methane.

280. It is a colorless gas, having no odor. Its density compared to air is 0.559, or compared to hydrogen, 8 : this corresponds to a molecular weight of 16, as is indicated by the formula, CH^4. It is a combustible gas, and burns with a yellow flame. It forms an explosive mixture with air or oxygen, and this mixture is often unfortunately formed in the galleries of coal-mines, for methane is the *fire-damp* of the miners. It exists under strong pressure in the coal-beds, and escapes when these beds are cut into by the miners.

We have already learned that a certain temperature is necessary for combustion, as indeed for all chemical action, and a gas cannot continue burning when its flame is cooled below the igniting point. When a flame is inserted in a tube, not too wide, it is extinguished, because the walls of the tube cool it. For this reason the flame does not run down the tube of a good Bunsen burner, although the combustible gas is mixed with air. A piece of wire gauze may be regarded as composed of a large number of fine, short tubes, and wire gauze will prevent the passage of flame. The fineness of the gauze required will depend on the igniting point of the gas or vapor, and, as this temperature is lower, the gauze must be finer. We may depress a piece of wire gauze in the flame of a Bunsen burner or a lamp, and the flame is kept below the gauze until the latter is heated to the temperature required for the combustion of the gas.

FIG. 85.

Yet the combustible gas passes through, and we may light it above the gauze : in the same manner we may hold the gauze a short distance above the burner in the escaping but unlighted gas, and we may ignite the gas above the gauze ; the flame does not, however, pass below until the gauze becomes heated as before (Fig. 85). These principles are applied in the miners' safety-lamp, which is practically a lamp so arranged that air can pass to the flame and the burned gases escape only through the meshes of fine wire gauze (Fig. 86). For better illumination, that part of the gauze immediately around the flame is usually replaced by thick glass. The explosive gases may enter this lamp, and may burn inside, but the flame cannot pass through unless the gauze become highly heated. In most countries it is unlawful to continue working galleries containing explosive gases until those gases are removed by ventilation. The safety-lamp affords a means of detecting the presence of very small quantities of such gases without danger of exploding them. We pass a very little illuminating gas or some of our methane into an inverted jar, and thoroughly mix it with the air in the jar by moving a roll of paper around in it. We now push up into the jar a lighted wax taper, the end of which projects just beyond a small glass tube slipped over it, so that the flame is quite small. We see that this small flame is surmounted by a pale and tremulous bluish cap (Fig. 87): this is owing to the combustion of the mixture of gas and air immediately around the flame, but there is so little of the combustible gas present that the

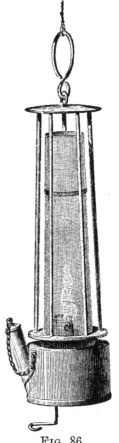

FIG. 86.

heat produced by its combustion immediately around the flame is not sufficient to carry the combustion throughout the whole mixture; otherwise there would be an explosion. By looking at the flame in his safety-lamp, the miner can tell by the presence or absence of this bluish cap whether any fire-damp be present, and, if so, whether there be sufficient to indicate danger of explosion.

281. Methane is one of the products of the putrefaction of vegetable matters in presence of water. It is formed by the decomposition of such substances in the muddy bottoms of ponds and rivers, and rises in bubbles through the water when this mud

m

is stirred: it often collects under the ice in winter, and will escape and burn with a pale flame when the ice is pierced and the gas lighted. Because of its formation in these localities, methane is often called *marsh gas.*

FIG. 87.

282. The composition of methane shows us that the carbon atom is tetratomic; it has the combining power of four atoms of hydrogen. We have already learned that chlorine has an energetic affinity for hydrogen, and that it will remove this element from many hydrogen compounds. When chlorine is mixed with methane, and the mixture is exposed to light, the chlorine removes the hydrogen from the methane, and hydrochloric acid is formed, but an atom of chlorine takes the place of every atom of hydrogen so removed. We may consider that there is a double decomposition between the chlorine molecules and the methane molecules, and this decomposition may continue until all the hydrogen atoms of the methane are replaced by chlorine.

$$CH^4 + Cl^2 = CH^3Cl + HCl$$
$$CH^4 + 2Cl^2 = CH^2Cl^2 + 2HCl$$
$$CH^4 + 3Cl^2 = CHCl^3 + 3HCl$$
$$CH^4 + 4Cl^2 = CCl^4 + 4HCl$$

All these compounds of carbon with chlorine and hydrogen may be obtained in this manner. Their compositions are a still further evidence that the carbon atom is tetratomic. The hydrogen atoms of methane may also be replaced by the monatomic atoms of bromine and iodine, producing compounds precisely similar to those formed by chlorine.

One of the substances so formed has the composition CH^3I; it is called *methyl iodide,* and the compound CH^3Cl is called *methyl* chloride. We may consider that the group of atoms CH^3 acts like a single atom of potassium in potassium chloride; and when we have learned that it may take part in double decompositions, leaving one molecule and entering another without change, we shall see that it is a radical; it is called *methyl.*

283. Methyl iodide, CH^3I, is a colorless liquid. When it is sealed up in strong glass tubes containing some zinc, and the tubes are heated for a time to about 150°, the zinc takes away the iodine from the methyl iodide, and zinc iodide, ZnI^2, is formed. When the tubes are carefully opened, they are found to contain a gas to which both analysis and density assign the composition C^2H^6. How must the atoms be related in a molecule of this gas? Are the carbon atoms still tetratomic? How has the

gas been formed? We must believe that when two atoms of iodine are removed from two molecules of methyl iodide the two monatomic methyl groups, CH^3, combine together; that in a molecule of the gas, C^2H^6, the two carbon atoms, each with its three hydrogen atoms, like three satellites, form a perfect system. We can represent this relation by our formulæ.

$$CH^3I \quad + \quad Zn \quad + \quad ICH^3 \quad = \quad ZnI^2 \quad + \quad H^3C\text{-}CH^3$$
Methyl iodide.   Zinc.   Methyl iodide.   Zinc iodide.   Ethane.

Then in this gas, C^2H^6, which is called ethane, the affinity of the carbon atoms must be satisfied partly by their combination together, and partly by their combination with hydrogen.

284. By the action of chlorine on ethane, the hydrogen of that gas may be replaced by chlorine atoms, and compounds may also be obtained in which the replacement is by iodine atoms. When only one of the hydrogen atoms is so replaced, the compound C^2H^5I is formed. We consider that this contains the radical C^2H^5, which is called *ethyl*, and the molecule C^2H^5I is called *ethyl iodide*. Since all the atoms of hydrogen in a molecule of ethane must have the same relations to the carbon atom around which they move, and also to the other carbon atom, it is a matter of indifference which one we suppose to be replaced by the iodine atom. When ethyl iodide and methyl iodide, in the proportions required for the same number of molecules of each, are heated with zinc in sealed tubes, a reaction takes place just as in the case of zinc and methyl iodide alone. That is, both iodine atoms are removed, and we may say either that the iodine of ethyl iodide is replaced by the group methyl, CH^3, or that the iodine of methyl iodide is replaced by the radical ethyl, C^2H^5. A gas called propane, C^3H^8, is then formed.

$$C^2H^5I \quad + \quad Zn \quad + \quad ICH^3 \quad = \quad ZnI^2 \quad + \quad C^2H^5\text{-}CH^3$$
Ethyl iodide.   Methyl iodide.   Zinc iodide.   Propane.

We find, then, that the atoms of carbon are able to combine together; that they form complex systems in which each carbon atom is accompanied by atoms of hydrogen or some other element. As we have done before, we may compare the carbon atoms to stars or suns which revolve around each other; each sun is accompanied by its own planets, and we shall presently see that each of the planets may have its satellites.

By reason of the property of combination between its own atoms, a property which is not possessed in the same degree by the atoms of any other element, carbon forms an almost infinite number of compounds. These compounds differ from those of the other elements in this respect :—while any other element forms a few compounds with nearly all other elements, carbon forms innumerable compounds containing very few of the other elements. The more numerous of the carbon compounds contain only carbon, hydrogen, oxygen, and nitrogen, but all of the other elements may, under proper conditions, be made to form part of these compounds. The carbon compounds have been commonly called organic compounds.

285. The compounds containing carbon and hydrogen only, are called hydrocarbons; we have just studied three of them, and in the molecules of each of these the combining power of the carbon atoms is completely exhausted. We may express in detail the atomic relations of the three.

$$
\begin{array}{ccc}
\text{H} & \text{H H} & \text{H H H} \\
| & | \ | & | \ | \ | \\
\text{H--C--H} & \text{H--C--C--H} & \text{H--C--C--C--H} \\
| & | \ | & | \ | \ | \\
\text{H} & \text{H H} & \text{H H H} \\
\text{Methane.} & \text{Ethane.} & \text{Propane.}
\end{array}
$$

The union of the carbon atoms together does not stop with propane, for in turn one of its hydrogen atoms may be replaced by a methyl group, and the hydrocarbon, C^4H^{10}, is the result. In the same manner this may be converted into C^5H^{12}, and a whole series of *saturated* hydrocarbons has been obtained. When we examine the composition of the members of this series, we see that each contains two more than twice as many atoms of hydrogen as it does of carbon. We may express the composition of any member of the series by the general formula C^nH^{2n+2}, n representing the number of carbon atoms in the molecule. The names of these compounds end in *ane*, and after the fourth member, the prefix indicates the number of carbon atoms in a molecule.

CH^4,	Methane.	C^5H^{12},	Pentane.
C^2H^6,	Ethane.	C^6H^{14},	Hexane.
C^3H^8,	Propane.	C^7H^{16},	Heptane.
C^4H^{10},	Butane.	C^8H^{18},	Octane.

The first five are gases at ordinary temperatures; the others are liquids of which the boiling points are higher as the number of carbon atoms in the molecule increases, until, when this number reaches sixteen, the compounds are solid at ordinary temperatures. Ordinary paraffin is a mixture of the solid members of the series; its name, meaning poor affinity, indicates that it does not readily enter into chemical reactions, and, since this property is common to all of the saturated hydrocarbons, the series $C^n H^{2n+2}$ is often called the paraffin series.

We see that each member of this series contains one atom of carbon and two atoms of hydrogen more than the preceding. Compounds which thus differ from each other by CH^2, or a multiple of that symbol, and which have the same general chemical properties, are said to be *homologous*, and to form a homologous series.

LESSON XXXV.

COMPOUNDS OF CARBON AND HYDROGEN (2).

286. **Petroleum.**—Petroleum, or rock-oil, as the name signifies, has been known from very early history, but it has been marvellously abundant in commerce only since 1859, when it was found that the oil would flow from wells bored into the rock in Northwestern Pennsylvania. The oil usually occurs in a loose, coarse sandstone into which it has drained from its source in other rocks. That source is still a matter of uncertainty, but the oil has doubtless been formed by the decomposition of vegetable and perhaps animal matters, long buried in the earth. The depth to which the wells must be sunk varies with each locality; sometimes it is only a few feet; sometimes it may be two or three thousand feet. Sometimes the oil begins to flow as soon as the oil-bearing rock is penetrated, but more usually the interior pressure is not strong enough to raise the oil, and a pump must then be employed. Petroleum is widely distributed; there are few portions of the world where it is not found, and there are immense oil-fields in Austria and Russia.

Crude petroleum varies in color from pale yellow to almost black; it usually has a greenish tint. It is sometimes quite fluid, sometimes thick like molasses. Its density is comprised between 0.75 and 0.92. It is a mixture of a large number of hydrocarbons,

16

all of which, with trifling exceptions, belong to the class of par-
affins which we have just studied.   Indeed, all the saturated
hydrocarbons, from CH^4 up to $C^{16}H^{34}$, have been separated from
coal-oil, as petroleum is commonly called.   These compounds have
different boiling points, and by slowly raising the temperature the
most volatile pass off and can be condensed first.   In the manufacture, the oil is slowly heated to about 70°, and the portion
which distils over is called *naphtha;* the temperature is then
raised to about 150°, and the liquid condensed up to that point
is *benzine:* between 150° and 280°, *kerosene,* or illuminating oil,
distils over, and that portion which passes between 280° and 400°
is paraffin oil or lubricating oil.   Much paraffin distils towards
the close of the operation, and a residue of coke remains in the
retort.

Naphtha has a density of about 0.65, and, when purified from
its most volatile constituents, forms *gasoline,* used in some gas-
machines.   Air is blown through the gasoline, and becomes charged
with sufficient of the vapor of the volatile hydrocarbons to burn
with an illuminating flame.   Benzine has a density of about 0.702,
and boils at about 148°.   It is used for dissolving oils and fats, and
instead of turpentine for mixing with paints.

Kerosene should contain no product whose boiling point is
below 150°, for the vapors of the more volatile hydrocarbons form
dangerously explosive mixtures with air.   The fire-test by which
the safety of the oil is determined, is made by slowly heating the
oil in a little dish on a water-bath, carefully observing by means
of a thermometer the temperature at which inflammable vapors
are given off and the temperature at which the oil takes fire.
A lighted match is passed rapidly over the oil, about half a centimetre from its surface; when the vapor burns with a little
flash, the thermometer marks the *flashing-point.*   A few degrees
above this, the oil itself takes fire.   The flashing-point should not
be below 60°, and the burning-point not below 65°.

287. **Paraffin.**—The name paraffin is commonly applied to
that product of the distillation of petroleum which solidifies on
cooling: it is also a product of the destructive distillation of peat

and some kinds of coal. When the last liquid portions of the distillate of petroleum are cooled by ice, a considerable quantity of paraffin separates. When purified, paraffin is a colorless, transparent, or translucent mass. It is a mixture of several members of the series of saturated hydrocarbons. Accordingly as it has been prepared and purified, its melting-point varies from 45° to 65°. It makes excellent candles.

288. We have now learned something about one class of hydrocarbons, a class in which the carbon atoms cannot combine with any other atoms unless they separate from each other. It is worthy of notice that they do not separate from each other except by the action of the most energetic agents: on the contrary, these carbon atoms remain combined, and, with as many of their accompanying hydrogen atoms as we permit to remain with them, constitute fixed and definite radicals, which act exactly like the atoms of elements having the same combining powers. We have noticed two of these radicals, methyl and ethyl. In order that there may be a uniformity of names for these complex groups, chemists have agreed to retain the first syllable of the name of the saturated hydrocarbon, in the names of all compounds derived from that hydrocarbon. The termination in *yl* has been selected for the radicals which we consider are formed by the removal of one atom of hydrogen from a saturated hydrocarbon, and then the first word of the name of a compound will show us the hydrocarbon radical in the molecule, and the last word must indicate the atom or group of atoms combined with that radical. We may then understand the composition of the following bodies:

CH^4, Methane.	CH^3Br,	Methyl bromide.	$CH^3.OH$,	Methyl hydrate.
C^2H^6, Ethane.	C^2H^5Cl,	Ethyl chloride.	$C^2H^5.OH$,	Ethyl hydrate.
C^3H^8, Propane.	$C^3H^7NH^2$,	Propyl amine.	$(C^3H^7)^2O$,	Propyl oxide.

When we have once acquired definite ideas of what is meant by a radical, that it is a group which acts precisely as an atom, having continually the same combining power or atomicity, leaving one molecule and entering another as a distinct existence; then the structure of these complex molecules becomes perfectly intelligible, and we need only be acquainted with the radicals concerned in order to be able at once to interpret, by our system of atomic groupings, the relations of the atoms in the molecule of any compound.

289. Unsaturated Hydrocarbons.

We have mixed in a glass flask some alcohol with four times its weight of strong sulphuric acid, and, as this mixture sometimes froths very much when we heat it, we have put in enough sand to absorb the liquid almost entirely. After fitting to our flask a cork through which pass a delivery-tube, and a safety-tube in which we put a little mercury or some sulphuric acid, we heat it on a sand-bath. A gas is dis-engaged, and we may collect it in jars over the pneumatic trough.

290. ETHYLENE, C^2H^4.—The gas which we have prepared is a hydrocarbon. It is colorless and almost odorless: its density compared to air is 0.9784, or compared to hydrogen, 14. Analysis shows that it contains carbon and hydrogen in the proportion of one atom of the first to two atoms of the second, and its density shows that its molecule must contain two atoms of carbon and four of hydrogen. Its composition is, then, C^2H^4: it is called ethylene. It burns with a brilliant flame.

Into a jar of this gas we pour a little bromine, and cause it to flow over the sides of the jar: the color of the bromine disappears, and drops of an oily liquid are formed. This liquid has a pleasant odor, very different from the suffocating vapor of the bromine. The ethylene has combined with the bromine and formed this liquid, which is called ethylene bromide. The vapor-density and analysis of the compound assign to its molecule the composition $C^2H^4Br^2$. Evidently if the molecule C^2H^4 can combine directly with two atoms of bromine, it must be a diatomic molecule, capable of manifesting the combining power of two atoms of hydrogen. Let us study the reaction by which ethylene is formed: alcohol is ethyl hydrate, $C^2H^5.OH$: sulphuric acid, by its strong affinity for water, converts it into $H^2O + C^2H^4$. Then, in losing the mon-atomic hydroxyl group and an atom of hydrogen, the carbon atoms of alcohol must recover the combining powers of two atoms of hydrogen: this combining power is manifested in the combination of ethylene with bromine, chlorine, etc. If two atoms of hydrogen were removed from a molecule of methane, CH^4, the remaining group, CH^2, would be diatomic, and we believe that a molecule of ethylene gas is formed by the union of two such diatomic groups, and is expressed by the formula $CH^2=CH^2$; but these atoms then possess more energy than when combined in the gas ethane, CH^3-CH^3, and may develop that energy and enter into direct combination with bromine, forming ethylene bromide, CH^2Br-CH^2Br.

ETHYLENE CHLORIDE, CH^2Cl-CH^2Cl, is formed when equal volumes of chlorine and ethylene are mixed in diffuse daylight.

It is a somewhat oily liquid, and from this character ethylene was first called olefiant gas. It boils at 82°. It boils at 82°.

. ETHLYENE BROMIDE, CH^2Br-CH^2Br, is made by passing ethylene gas into cooled bromine. It boils at 131°.

291. We have seen that chlorine is capable of replacing the hydrogen of ethane, C^2H^6, atom for atom. From the products of this reaction we can by careful operations separate two liquids having the composition $C^2H^4Cl^2$, but having entirely different properties. These compounds are isomeric, and we may understand their isomerism when we see that both atoms of chlorine may replace hydrogen atoms which are in relation to the same atom of carbon, forming the molecule CH^3-CHCl^2; or each may replace an atom of hydrogen from a group CH^3; the compound formed in the latter case would of course be ethylene chloride.

292. Ethylene is only the first member of a long series of hydrocarbons which we may consider are derived from it by the replacement of one or more of its hydrogen atoms by the monatomic hydrocarbon radicals which we have already studied. Each of the compounds so formed is diatomic: it will combine directly with two atoms of chlorine or bromine, and may be made to combine with two monatomic radicals or with one diatomic radical. The names of these diatomic hydrocarbons are made to correspond with the saturated hydrocarbons, from which we may consider they are derived by the removal of two atoms of hydrogen, but the *ane* of the name is changed to *ylene*. Ethylene corresponds to ethane, butylene corresponds to butane. We have here our second series of homologous compounds, each differing from the next by CH^2.

C^2H^4, Ethylene.	C^5H^{10}, Amylene or pentylene.
C^3H^6, Propylene.	C^6H^{12}, Hexylene.
C^4H^8, Butylene.	C^7H^{14}, Heptylene, etc.

On examination, we notice that each molecule contains twice as many atoms of hydrogen as of carbon; the general formula for the series is C^nH^{2n}. The proportion of hydrogen and carbon is the same in each member of the series, but the molecular weights, and consequently the number of atoms in the molecules, are not the same. Bodies of which the molecules contain the same atoms in the same proportion but in different numbers are said to be *polymeric*. All of these diatomic hydrocarbons are polymeric; the number of carbon atoms and hydrogen atoms in each is an exact multiple of CH^2. Because these hydrocarbons combine directly with chlorine and bromine, forming oily liquids, the series is often called the olefin series.

293. It has been said that we may consider these bodies as formed from ethylene by the replacement of hydrogen atoms by the monatomic radicals,

methyl, ethyl, etc. We must see that this replacement may yield many in-stances of isomerism. If one of the hydrogen atoms of ethylene be replaced by methyl, we obtain propylene.

$$CH^2{=}CH^2 \qquad\qquad CH^2{=}CH{-}CH^3$$
Ethylene.   Propylene.

By the replacement of two of the hydrogen atoms by methyl, we may obtain two different butylenes, according to the positions of the replaced hydro-gen atoms, and there is still a third butylene, formed by the replacement of one hydrogen atom by an ethyl group.

$$CH^2 \qquad\quad CH.CH^3 \qquad\quad C(CH^3)^2 \qquad\quad CH(C^2H^5)$$
$$CH^2 \qquad\quad CH.CH^3 \qquad\quad CH^2 \qquad\qquad CH^2$$
Ethylene.   (a) Dimethylethylene. (β) Dimethylethylene.   Ethylethylene.

All these hydrocarbons have been obtained and studied, and their names indicate the molecules from which they are derived and the radicals which are substituted for the hydrogen atoms in those molecules.

We can understand that after the third member of the series of saturated hydrocarbons, a similar isomerism must exist for those compounds also.

LESSON XXXVI.

COMPOUNDS OF CARBON AND HYDROGEN (3).—
ANALYSIS OF CARBON COMPOUNDS.

294. The tar which condenses during the distillation of bitu-minous coal for the manufacture of gas, is an exceedingly com-plex liquid, consisting principally of compounds of carbon and hydrogen. Some of these compounds are solid, some of them are volatile liquids. Since they boil at different temperatures, they can be separated from each other by a process called fractional distillation. When a mixture of liquids having different boiling points is distilled very slowly, the most volatile portions pass off first and may be condensed. In practice, the boiling is not con-ducted very slowly, but the mixed vapors of the substances are passed through a tube or pipe which is maintained at the boiling point of the most volatile constituent of the mixture: in this tube the liquids having higher boiling points are condensed, and flow back into the still, while the vapor of the more volatile liquid

passes on and is condensed separately. The most simple apparatus which we can employ in the laboratory is a rather wide tube on which a couple of bulbs are blown; this is placed vertically in the flask in which we boil the mixed liquid (Fig. 88). The lower part of the tube becomes heated to the temperature of the mixed vapor, which is between the boiling points of the liquids: as some of the most easily condensed vapor is cooled to the condensing point, and converted into a liquid, the temperature of the tube gradually falls as the vapors approach the upper portion, and by carefully regulating the boiling, only the most volatile liquid passes from the apparatus, and this is indicated by a thermometer of which the bulb is opposite the short side-tube.

Fig. 88.

295. **Benzol,** C^6H^6.—The most volatile constituent of coal-tar is a liquid called benzol. It freezes at 5.5°, and boils at 80.5°. It does not dissolve in water, but is soluble in alcohol and ether. It is very inflammable, and burns with a bright but smoky flame. The composition of its molecule is C^6H^6, and yet in most of its reactions it acts like a saturated hydrocarbon. We put a few crystals of iodine into some benzol in a glass flask, and pass chlorine through the liquid; hydrochloric acid gas is given off, and the hydrogen atoms of the benzol are replaced by chlorine.

$$C^6H^6 + Cl^2 = C^6H^5Cl + HCl$$

The iodine only aids in the reaction by helping to break up the molecules of chlorine.

Evidently the molecular structure of benzol must be different from that of the other hydrocarbons which we have studied, and we can only account for its resemblance to the saturated hydrocarbons by supposing that its carbon atoms are differently related. Of several theories which have been proposed in order to explain the reactions of benzol, we need only consider one, which supposes that these atoms form a complex system, in which each carbon atom is combined with three other carbon atoms, so leaving one atomicity of each free to combine with an atom of hydrogen or other element. We may, by our formulæ, represent this arrangement in the form of a triangular prism, a carbon atom being placed at each angle.

296. All the hydrogen atoms of benzol may be replaced by other atoms or radicals, and, when more than one is so replaced, we have interesting isomeric compounds, the isomerism depending on the relations of the carbon atoms whose hydrogen atoms are affected. Let us suppose, for example, that two hydrogen atoms are replaced by two chlorine atoms: experiment has shown that three compounds may then be formed, having precisely the same composition, but different properties.

We can interpret this by our theory and our representation of the molecule, by considering that while one atom of chlorine always occupies the same place, the position of the other varies.

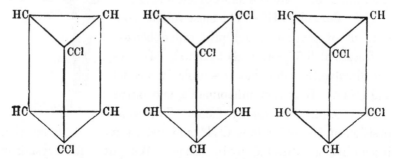

While our time will not permit the further development of these ideas, we may say that theory indicates only three positions in which the two chlorine atoms can form entirely different molecules. By different methods chemists have always succeeded in producing three isomeric compounds in which two atoms of hydrogen of benzol are replaced by the same atoms or radicals, but they have never been able to obtain more than three such compounds.

297. There are many hydrocarbons which we believe to be derived from benzol by the replacement of its hydrogen atoms by radicals, such as methyl, ethyl, etc. Some of these have been obtained by methods which allow no doubt as to their constitution; others have not yet been so formed, but certain of their chemical reactions seem to show that they also are derived from benzol. These hydrocarbons and many of the bodies derived

from them have peculiar aromatic odors, and for this reason the whole series of compounds which are considered as benzol derivatives is commonly called the *aromatic series*. Of these compounds we can consider only a few.

298. **Methyl-benzol,** $C^6H^5.CH^3$, is generally called toluol, because it was first derived from tolu balsam. It is now obtained from coal-tar, and usually constitutes a considerable proportion of the benzol of commerce. It is a colorless liquid, boiling at 111°.

We can understand that there may be four isomeric compounds formed by the replacement of a single hydrogen atom of methyl-benzol, for that replacement may affect either a hydrogen atom in one of three places in the benzol group, C^6H^5, or a hydrogen atom of the radical methyl, CH^3.

The three isomeric dimethyl benzols, $C^6H^4(CH^3)^2$, are called xylols: two are liquids, and one is a solid. Isomeric with them is also ethyl-benzol, $C^6H^5.C^2H^5$.

299. **Oil of Turpentine,** $C^{10}H^{16}$, is a derivative of benzol, and it is isomeric with a large number of essential oils. The oils of lemon, orange, bergamot, juniper, lavender, and many others, all appear to have the same molecular composition, and we must believe that their differences are due to a different arrangement of the atoms constituting their molecules. These oils are obtained by distilling with water the leaves or other parts of the plant containing them. It is true that the boiling point of each of these oils is much higher than that of water, but the steam of the water readily carries over the oil. The condensed liquid then separates into two layers, the lower being water, and the upper and lighter being the essential oil. Oil of turpentine is so made by distilling with water the crude turpentine which flows from incisions made in certain species of pine-trees. There are several varieties of this oil, which differ according to the species of pine-tree which furnishes them. The density is about 0.87, and they boil at about 156°. Oil of turpentine and most of the essential oils slowly absorb oxygen from the air, and are converted into various resins.

300. **Naphthaline,** $C^{10}H^8$, is a solid hydrocarbon derived from coal-tar. It usually occurs as pearly scales, melting at 79°, and boiling at 218°. It does not dissolve in water, and but slightly in cold alcohol. It is soluble in boiling alcohol, and crystallizes

when the solution cools. It is employed for the manufacture of numerous beautiful dye-stuffs, analogous to the aniline dyes, which we will presently study.

It will be noticed that naphthaline is isomeric with the three xylols and with ethyl benzol, and makes the fifth hydrocarbon of the composition $C^{10}H^8$.

301. Anthracene, $C^{14}H^{10}$, is one of the least volatile hydrocarbons obtained from coal-tar. When pure, it forms beautiful transparent prisms, which melt at 213° and boil at about 360°. It is employed for the manufacture of alizarin, a red coloring matter which was until within a few years obtained only from madder. The ability to produce this dye-stuff by purely chemical processes has permitted large areas of land which were formerly devoted to the cultivation of the madder-plant to be used for raising grain. Besides this, chemists have been able to prepare from this same alizarin valuable dye-stuffs of other colors, and there is now a whole series of anthracene coloring matters.

ANALYSIS OF CARBON COMPOUNDS.

302. The proportions in which the elements exist in any carbon compound are determined by elementary analysis. If the compound contains other elements than carbon, hydrogen, and oxygen, its analysis requires several operations: if only these three elements be present, the carbon and hydrogen are determined by one operation, and the quantity of oxygen is the difference between the sum of the carbon and hydrogen and the total weight of the substance analyzed. The analysis is conducted by mixing a weighed quantity of the substance with pure and dry cupric oxide in a long glass tube, one end of which is drawn out to a fine point and sealed. The other end is connected with a small U-tube containing pumice-stone wet with sulphuric acid, and the U-tube is connected with a bulbed tube containing a solution of potassium hydrate (Fig. 89). The tube is heated to redness in a long tube-furnace, and the oxygen of the cupric oxide converts the hydrogen of the carbon compound into water, while the carbon is burned into carbon dioxide. The water is absorbed in the tube containing the pumice and sulphuric acid, while the carbon dioxide is absorbed by the potassium hydrate. Towards the close of the operation, a caoutchouc tube connected with an oxygen gas-holder is slipped over the drawn-out point of the combustion-tube; the oxygen is turned on, and the point is broken off by pinching the end of the tube. A current of pure dry oxygen is then passed through the red-hot tube, and all traces of carbon dioxide and watery vapor are forced through the absorption-tubes; at the same time any unburned carbon is completely consumed, and the copper from which oxygen has been removed is again converted into cupric oxide

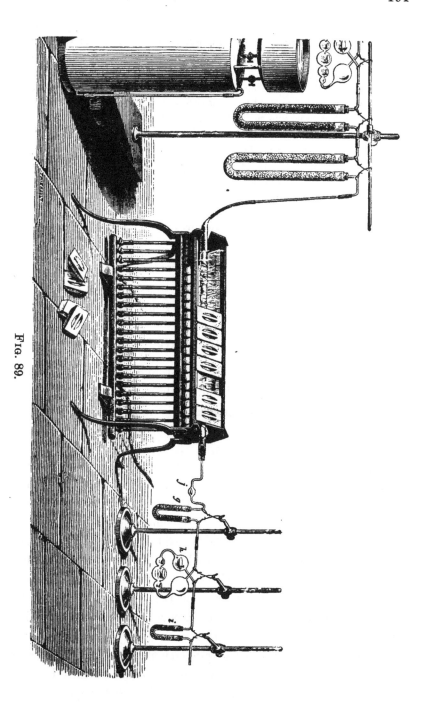

FIG. 89.

for another operation. The increased weight of the U-tube (j and g), in which water has been absorbed, is due to the water, and one-ninth of the increase will represent the quantity of carbon in the amount of substance analyzed. The increased weight of the bulbed tube (h) is due to carbon dioxide, and $\frac{12}{44}$, or $\frac{3}{11}$, of this increase will give us the quantity of carbon which we wish to determine. Since the current of oxygen would carry a little vapor of water out of the bulbed tube, and so diminish its weight, a small tube (i) containing pumice and sulphuric acid is attached, and in this the vapor is retained; this tube is always weighed with the potash bulbs.

Having determined the proportions of all the elements in a compound, its molecular weight is calculated from its vapor-density, or, if this be not possible, by indirect methods. Knowing the molecular weight and the proportion of each element present, it is very easy to fix the chemical formula expressing the composition of the molecule.

LESSON XXXVII.

ALCOHOLS (1).

303. When the iodide of a radical like methyl or ethyl is heated with silver oxide and water, silver iodide is formed, and the iodine of the carbon compound is replaced by a hydroxyl group.

$$2CH^3I \quad + \quad Ag^2O \quad + \quad H^2O \quad = \quad 2AgI \quad + \quad 2CH^3.OH$$
Methyl iodide.   Silver oxide.   Silver iodide.   Methyl hydrate.

A hydrate of the hydrocarbon radical is so formed, and these hydrates constitute what are called the alcohols. We must study some of the more important of these compounds.

304. **Methyl Alcohol, $CH^3.OH$.**—The liquid which condenses during the manufacture of charcoal (§ 226) contains small quantities of a volatile liquid which can be separated by careful fractional distillation. This liquid is usually sold under the name methylene or wood-spirit. It is impure methyl alcohol, and is used for the manufacture of varnishes, and for dissolving fats and oils. Methyl alcohol has the property of forming with calcium chloride a crystalline compound, and is usually purified by saturating the wood-spirit with calcium chloride, and evaporating the

solution by a gentle heat until it crystallizes. The crystals are dissolved in water; when their solution is boiled, the compound of methyl hydrate and calcium chloride is decomposed, and the methyl alcohol can be separated by fractional distillation.

Pure methyl alcohol is a colorless liquid, of an odor exactly like that of common alcohol. Its density at $0°$ is 0.814, and it boils at $66.5°$. It mixes in all proportions with water and with ordinary alcohol. It is inflammable, and burns with an almost colorless flame. We throw a piece of sodium into methyl alcohol : hydrogen is given off, and the metal dissolves : an atom of sodium has replaced the hydrogen atom of the group hydroxyl, and sodium methylate is formed. The reaction is precisely like that which yields sodium hydrate in the reaction of sodium with water.

$$2H\text{-}O\text{-}H \quad + \quad Na^2 \quad = \quad 2NaOH \quad + \quad H^2$$
$$2CH^3\text{-}O\text{-}H \quad + \quad Na^2 \quad = \quad 2CH^3\text{-}ONa \quad + \quad H^2$$
Methyl alcohol.   Sodium methylate.

When methyl alcohol is oxidized slowly, an atom of oxygen replaces two hydrogen atoms of the methyl group, CH^3, and formic acid results.

$$CH^3.OH \quad + \quad O^2 \quad = \quad H^2O \quad + \quad CHO.OH$$
Methyl hydrate.   Formic acid.

305. **Ethyl Alcohol,** $C^2H^5.OH$.—When ethylene gas is passed into strong hydriodic acid, direct combination takes place, and ethyl iodide is formed.

$$C^2H^4 \quad + \quad HI \quad = \quad C^2H^5I$$

When this ethyl iodide is heated with potassium hydrate solution, a double decomposition takes place, yielding potassium iodide and ethyl alcohol.

$$C^2H^5I \quad + \quad KOH \quad = \quad KI \quad + \quad C^2H^5.OH$$

However, ethyl hydrate, which is ordinary alcohol, is manufactured by a peculiar decomposition of glucose, or some substance having the same composition as glucose. This decomposition is brought about by a minute organism which lives and multiplies by converting the glucose into carbon dioxide and alcohol. A decomposition due to such an organized being is called a fermentation, and the organism is called a ferment. The molecule of glucose is

expressed by the formula $C^6H^{12}O^6$, and, although small quantities
of other substances are produced during the fermentation, which
is caused by the yeast-plant and is called the alcoholic fermentation, the general change may be represented by the equation

$$C^6H^{12}O^6 \;=\; 2C^2H^5.OH \;+\; 2CO^2$$

For the manufacture of alcohol, the product of the fermentation
is distilled, and the alcohol so separated from the water. However,
the best apparatus does not give alcohol stronger than about ninety-
five per cent. Pure or, as it is commonly called, *absolute* alcohol
is made by putting quick-lime into the strongest alcohol of commerce, and distilling the mixture after it has stood several days.
By reason of its strong affinity for water, the lime then retains all
of that liquid.

Pure alcohol is a colorless liquid, having a faint but pleasant
odor. Its density at 0° is 0.8095, and it boils at 78.4°. It mixes
with water in all proportions, and the mixture becomes slightly
warm and contracts in volume. Alcohol dissolves many substances
which are insoluble in water; among these are iodine, the essential
oils, fats, and resins. The *spirits* of the pharmacies, such as spirits
of ammonia, are solutions of volatile substances in alcohol; *tinctures* are similar solutions of non-volatile substances.

Alcohol is combustible, and burns with a pale flame, the products of the combustion being carbon dioxide and water.

By the slow oxidation of alcohol, acetic acid and a volatile
liquid called aldehyde are formed; acetic acid by the replacement
of two atoms of hydrogen of the ethyl group by an atom of oxygen, and aldehyde by the replacement of the hydroxyl group and
one atom of hydrogen by an atom of oxygen. Water is of course
formed in both cases.

$$CH^3-CH^2.OH \;+\; O \;=\; CH^3-CHO \;+\; H^2O$$
Alcohol.   Aldehyde.

$$CH^3-CH^2.OH \;+\; O^2 \;=\; CH^3-CO.OH \;+\; H^2O$$
Alcohol.   Acetic acid.

The slow oxidation of alcohol may be made to develop considerable heat. Over a little alcohol in a beaker we suspend a coil
of platinum wire which we have previously heated to·redness

(Fig. 90). The wire becomes bright red, and will continue to glow as long as sufficient air and alcohol vapor come in contact with the platinum. At the temperature of the red-hot wire the alcohol vapor is fully oxidized, but if we remove it, and allow it to cool slightly, and then withdraw it before it becomes bright, we may notice the peculiar odor developed in the beaker. This is in part due to the formation of aldehyde.

FIG. 90.

306. The reaction of alcohol with solutions of certain metals in nitric acid yields a class of bodies called the fulminates. Fulminating mercury, which is used for charging percussion-caps, may be prepared by dissolving about two grammes of mercury in fifteen cubic centimetres of strong nitric acid contained in a rather large flask or beaker. The reaction is aided by a gentle heat, and as soon as all the mercury has disappeared, the vessel is removed from the proximity of flame, and twenty cubic centimetres of alcohol are added. A violent reaction takes place, dense, white, poisonous vapors are disengaged, and fulminate of mercury is deposited as a light-gray powder. When the effervescence has ceased, the vessel is filled with water, and the acid liquid is poured off: the mercuric fulminate is washed by decantation, until the water no longer becomes acid. It is then collected on a small filter, and dried by exposure to the air. The reaction by which this compound is formed is very complicated, but the composition of mercuric fulminate is expressed by the formula $HgC^2N^2O^2$, and its molecule is believed to represent methane, CH^4, in which the four atoms of hydrogen are replaced by a group NO^2, a cyanogen group, and a diatomic atom of mercury,

CH^4, Methane. $C(NO^2)(CN)Hg$, Fulminate of mercury.

Fulminate of mercury explodes violently by friction or percussion, and should be kept in loosely-corked bottles, lest it be exploded by the friction of a glass stopper. It explodes also at a temperature of about 180°. Although this body is so exceedingly explosive that it would burst a gun-barrel in which it was detonated, the expansive force of the gases produced is much inferior to that of those disengaged by gunpowder, and it could not be used for projectile effects.

307. ALCOHOLIC BEVERAGES are products of the fermentation of substances containing glucose or some body capable of being converted into glucose. In the manufacture of *wine*, which is the natural beverage of those countries in whose climates the wine-grapes flourish, the glucose is derived from the juice of the grape: the ferment also is natural to the grape, for it is developed from the albumen like matter of the pulp. Since the alcoholic fermen-

tation is a transformation of glucose, and no air is necessary for the change, this fermentation may continue in closed vessels; in *sparkling wines* or champagnes part of the fermentation takes place in the bottle, and the carbon dioxide formed is dissolved in the liquid under pressure. All the carbon dioxide has escaped from *still wines.* The fermentation of apple-juice and the juices of other fruits, which yields cider and the various fruit-wines, is quite similar to the fermentation of grape-juice. Wines contain from seven to twenty per cent. of alcohol.

Beer, ale, and *porter* are produced from grain, preferably from barley. Grain contains no glucose, but during the sprouting of the grain, a body precisely similar to glucose and having the same composition is formed. It is called *maltose.* The barley is therefore moistened and kept at a temperature of about 15° until a sprout as long as the grain is formed. The sprouting is then arrested by heating the grain, which is now called *malt,* to about 50°, after which the dry malt is ground to a coarse powder and is ready for brewing. It is then cooked for several hours with water at a temperature of 50° or 60°, and all the maltose and much of the nutritious matter of the malt are dissolved : the liquid thus formed is heated with hops to impart an aromatic flavor, and is then rapidly cooled; after a little yeast is added, the wort is allowed to ferment at as low a temperature as possible, until in a few days beer or ale is obtained, according to the proportions of substances used.

Beer contains from two to five per cent. of alcohol, and ale a somewhat larger proportion, sometimes as high as ten per cent. As there is in this country no government inspection of malted liquors, beer is often adulterated by the substitution of various more or less injurious bitter substances for the hops, and of glucose for a part of the malt : glucose is not injurious, but it contains no nutritious matter, as is the case with malt.

Spirituous liquors are not natural products ; they are distilled from various fermented liquids, and are only dilute alcohol containing some flavoring matter. *Brandy* is distilled from wine ; *whiskey* from malted liquors of all kinds, derived from corn, rye, oats, and even potatoes; *rum* is distilled from fermented molasses from sugar-cane; gin is dilute alcohol flavored with the essential oil of juniper-berries. These liquids contain from forty to sixty per cent. of alcohol.

LESSON XXXVIII.

ALCOHOLS (2).

308. Propyl Alcohols, $C^3H^7.OH$.—A substance of this composition exists in very small proportion among the products of the alcoholic fermentation. It is a liquid, boiling at 98°. When we examine the composition of the hydrocarbon propane, we will notice that the three carbon atoms are not similarly related: two are related to one other carbon atom, but the third is related to both of the first two.

$$H-\overset{\overset{H}{|}}{C}-\overset{\overset{H}{|}}{C}-\overset{\overset{H}{|}}{C}-H$$

Chemists have discovered two propyl alcohols, and indeed two modifications of every compound containing the radical propyl, C^3H^7. We account for this in our theory by considering that in one of these alcohols the hydroxyl group replaces one of the hydrogen atoms of either of the two carbon atoms which are related to only one other carbon atom, and we see that all these atoms of hydrogen are similarly situated. The alcohol so formed is *normal propyl alcohol*, that which we have briefly considered. In the other alcohol of the same composition the hydroxyl group is joined to that carbon atom which is related to two others: it is called *isopropyl alcohol*, and boils at 86°. We see, then, that there are two propyl radicals; propyl, $CH^3–CH^2–CH^2$; and isopropyl, $CH(CH^3)^2$. Those alcohols in which the carbon atom which holds the hydroxyl group in the molecule is related to only one other atom of carbon, are called *primary alcohols*. Those in which the hydroxyl group is in

17*

relations with an atom of carbon which is related to two others, are called secondary alcohols.

309. Butyl Alcohols, $C^4H^9.OH$.—Chemists have succeeded in preparing four different butyl alcohols. They are all liquids, with the exception of one, which is a crystalline solid, in whose molecule we believe that the hydroxyl group is held by a carbon atom which acts as the centre of a system; around this atom are grouped three other atoms of carbon and their accompanying hydrogen atoms. It is $C(CH^3)^3OH$. It is called a tertiary alcohol, that being the name applied to those alcohols in which the carbon atom which brings the hydroxyl group into the molecule is related to three other carbon atoms.

310. Amyl Alcohols, $C^5H^{11}.OH$.—There are now known seven alcohols of this composition. Two of them exist in the oily residue which is left in the distillation of brandy and whiskey, and they are therefore products of the alcoholic fermentation of glucose. This oily residue is called fusel oil: it has a peculiar and not altogether pleasant odor, and, besides some ordinary alcohol which it still retains, is a mixture of propyl, butyl, and two amyl alcohols. The first two may be isolated from the mixture by careful fractional distillation, but for the separation of the two amyl alcohols chemical means must be employed. The crude fusel oil is a valuable solvent for many substances, to dissolve which ordinary alcohol would otherwise be required. Butyl and amyl alcohols do not mix in all proportions with water, as do ethyl and propyl alcohols.

We pour small quantities of methyl, ethyl, butyl alcohol from fusel oil, and amyl alcohol from fusel oil, into four separate plates, and light them. We find that the first is the most combustible, and that we can light the last only with difficulty. The flame of the methyl alcohol is almost colorless, but the brightness of the flames increases to the amyl alcohol, which burns with a bright light. The effect of an increased number of carbon atoms combined in the molecule is then to render the compound less volatile, and more difficult to inflame.

311. Glycols.—We have learned that when ethylene gas,

C^2H^4, is passed into bromine, a direct combination takes place, and ethylene bromide, $C^2H^4Br^2$, is formed; we have also acquired some idea of the molecule of this new compound. When ethylene bromide is boiled with a solution of potassium carbonate in water, carbon dioxide is given off, and, in addition to the potassium bromide which is formed, the liquid contains a new body.

$$C^2H^4Br^2 + K^2CO^3 + H^2O = C^2H^4(OH)^2 + 2KBr + CO^2$$

This new compound, $C^2H^4(OH)^2$, is formed by the replacement of the two bromine atoms of ethylene bromide by two hydroxyl groups. It is called ethylene alcohol, and after the potassium bromide has crystallized it can be separated by careful fractional distillation, and then forms a syrupy liquid having a sweet taste. Since it is a neutral body, and is a hydrate, it is called an alcohol, and it is a diatomic alcohol, because it contains two hydroxyl groups. It is the first member of a series of diatomic alcohols derived from the hydrocarbons of the series C^nH^{2n}. From its sweet taste, Wurtz, its discoverer, gave it the name *glycol*, and the diatomic alcohols are often called glycols.

312. **Glycerin,** $C^3H^5(OH)^3$.—The fats and fatty oils are complex compounds containing the radicals of certain carbon acids, called the fatty acids (§ 338), and the radical of the well-known substance glycerin. By the action of metallic hydrates on these compounds, atoms of metal replace the glycerin radical, which combines with the hydroxyl groups which were before in the metallic hydrate molecules. At high temperatures, steam acts precisely like the metallic hydrates, and glycerin is manufactured by distilling fats and oils in a current of superheated steam; that is, steam which has been passed through very hot pipes. The radical of glycerin having been found to be the group C^3H^5, we may represent the change thus:

$$C^3H^5(\text{fatty acid radical})^3 + 3HOH = C^3H^5(OH)^3 + 3H(\text{fatty acid radical}).$$

| Fat or Oil. | Water. | Glycerin. | Fatty acid. |

Glycerin is a colorless, syrupy liquid, having a sweet taste. Its density is about 1.28. It freezes at about 0°, and melts at about 7°. When it is heated, it boils at about 280°, but is partially decomposed, producing a very irritating odor of a substance called acrolein. It may be distilled in a vacuum or in a current of steam. It dissolves in all proportions of water and alcohol.

A molecule of glycerin contains three hydroxyl groups ; glycerin is then a triatomic alcohol. Each of these hydroxyl groups may be replaced by a monatomic atom or radical, and a large number of glycerin derivatives have been so formed.

313. NITROGLYCERIN, $C^3H^5(NO^3)^3$.—In a little beaker glass placed in ice-water, we have prepared a mixture of equal volumes of strong sulphuric and nitric acids; into this cold mixture we pour a few drops of glycerin, and, after stirring for a few moments, we throw the contents of the beaker into another glass nearly filled with cold water. A few drops of an oily liquid separate, and fall to the bottom of the glass; we pour off nearly all the water ; then, by means of a glass tube drawn out to a small opening, we remove a drop of the oil, and place it on the corner of a piece of paper ; when we light this, it burns with a bright flash. We now allow another small drop to fall on a nearly red-hot piece of sheet iron ; it explodes with a loud report. We place another small drop on an anvil, and when we strike it with a hammer there is another loud explosion. The oil is nitroglycerin ; it has been formed by the removal of three molecules of water from one molecule of glycerin and three molecules of nitric acid, and the union of the remaining groups, C^3H^5 and $3NO^3$.

$$C^3H^5(OH)^3 + 3HNO^3 = 3H^2O + C^3H^5(NO^3)^3$$

When nitroglycerin explodes, carbon dioxide and water are formed. The energy of the explosion is due to the energy of motion of the atoms in a molecule of nitroglycerin being greater than that energy in the carbon dioxide, water, and nitrogen, and during the explosion the excess of energy appears as heat and the energy necessary to convert the products into the gaseous state. Nitroglycerin is used in blasting operations, but it is usually mixed with a very fine sandy-earth, and the mixture is called dynamite. This is made into cartridges which are exploded by a fuse and detonating cap. Nitroglycerin is not a safe body to handle ; for experimental purposes we prepare only a few drops of it.

LESSON XXXIX.

SIMPLE ETHERS.

314. Oxides of Hydrocarbon Radicals.—In a glass flask we have cautiously mixed some methyl alcohol with about its own volume of strong sulphuric acid; after adapting a cork and straight tube to the flask, we heat it, and soon a colorless gas is disengaged. We light the gas, and it burns with a rather bright flame. (Fig. 91.) It is methyl oxide; the sulphuric acid has removed a molecule of water from two molecules of methyl alcohol, and the two methyl groups are held together by an atom of oxygen.

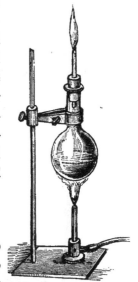

$$2CH^3.OH \quad = \quad (CH^3)^2O \quad + \quad H^2O$$
Methyl alcohol.   Methyl oxide.

The density of methyl oxide compared to hydrogen is 23. The gas is converted into a liquid at a temperature of —36°. It is soluble in water, alcohol, and ether.

FIG. 91.

In this compound we are again shown that the methyl group acts like an atom of hydrogen. It is a monatomic radical, and its composition is like that of water, excepting that, instead of two atoms joined by an atom of oxygen, two groups or systems are held to the oxygen atom.

$$H\text{-}O\text{-}H \qquad\qquad H^3C\text{-}O\text{-}CH^3$$
Water.   Methyl oxide.

The other monatomic radicals form similar oxides, and these oxides form part of a large class, called the simple ethers.

315. ETHYL OXIDE, $(C^2H^5)^2O$.—This compound, which is commonly called ether, may be formed by a number of reactions. In a strong glass tube we have sealed some ethyl iodide and silver oxide, and have heated the tube for several hours in boiling water. We now find that the black color of the silver oxide has changed to yellow, which is the color of silver iodide. A double decomposition has taken place, yielding silver iodide and ethyl oxide, which we may recognize by its odor and other properties when we cut open the tube.

$$Ag^2O \quad + \quad 2C^2H^5I \quad = \quad 2AgI \quad + \quad (C^2H^5)^2O$$

Silver oxide.　　Ethyl iodide.　　Silver iodide.　　Ethyl oxide.

In a rather large flask we have mixed some ninety per cent. alcohol with one and four-fifths times its weight of strong sulphuric acid. We adapt to this flask a cork having three holes; through one passes a thermometer (t, Fig. 92); another gives passage to a

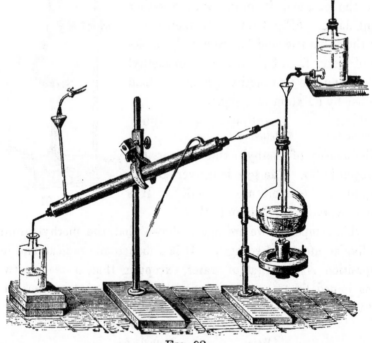

FIG. 92.

tube through which we may allow alcohol to flow from a reservoir, while to the third is fitted a delivery-tube connected with a condenser through which flows a stream of ice-water. We now heat the flask until the thermometer shows that the liquid into which the bulb dips has a temperature of 140°; then we regulate the flame so that the temperature may not rise further, and start a small stream of alcohol from the reservoir. This alcohol is quickly changed to ether, which, together with the water formed, distils and collects in the receiving-bottle.

The reaction which takes place in this operation is worth our study. When alcohol is mixed with strong sulphuric acid, water is formed, and an ethyl group is substituted for one atom of hydrogen of sulphuric acid, producing a compound called ethyl sulphuric acid.

$$C^2H^5.OH \;+\; H^2SO^4 \;=\; H^2O \;+\; (C^2H^5)HSO^4$$
$$\text{Alcohol.} \qquad\qquad\qquad\qquad \text{Ethylsulphuric acid.}$$

If this ethyl sulphuric acid is heated, it is converted into sulphuric acid and ethylene gas, C^2H^4; if it is boiled with water, it again yields alcohol and sulphuric acid; but if it is boiled with an additional quantity of alcohol, the result is sulphuric acid and ether.

$$(C^2H^5)HSO^4 \;+\; C^2H^5.OH \;=\; H^2SO^4 \;+\; (C^2H^5)^2O$$
$$\text{Ethylsulphuric acid.} \quad \text{Alcohol.} \qquad\qquad \text{Ether.}$$

If methyl alcohol be used instead of ethyl alcohol, a mixed oxide of methyl and ethyl is produced.

$$(C^2H^5)HSO^4 \;+\; CH^3.OH \;=\; H^2SO^4 \;+\; CH^3\text{-}O\text{-}C^2H^5$$

Ethyl oxide is a colorless, very mobile liquid, having a pleasant odor and a somewhat burning taste. Its density at $0°$ is 0.736; it boils at 34.5°. It will dissolve in nine times its weight of water, and one part of water will dissolve in thirty-six parts of ether. Ether dissolves small quantities of sulphur and phosphorus, and large proportions of bromine, iodine, fats, oils, and many other substances which are insoluble in water.

Ether is very inflammable, and its vapor forms an explosive mixture with air. We suspend a heated coil of platinum wire over a little ether in a beaker, as we made a similar experiment with alcohol, and the slow combustion of the ether vapor develops so much heat that the platinum wire becomes hot enough to inflame the ether.

The vapor of ether is very heavy,—2.564 compared to air. We

pour a little ether into a warm beaker, and then, holding the short end of a small siphon immediately above its surface, establish a cur-

rent of ether vapor by drawing out the air by the mouth; the heavy ether vapor continues to flow through the siphon, as would a liquid, and, when we light it, will burn as long as any ether remains in the beaker (Fig. 93).

The inhalation of ether vapor produces anæsthesia, and for this reason ether is largely employed as an anæsthetic in surgical operations.

FIG. 93.

The other monatomic hydrocarbon radicals form oxides corresponding to the oxides of methyl and ethyl: each of these contains two radicals related to one atom of oxygen. The diatomic hydrocarbons, ethylene and its homologues, form oxides containing one atom of oxygen and one molecule of the hydrocarbon, as in ethylene oxide, which may be obtained by heating ethylene bromide with silver oxide.

$$\begin{array}{c} CH^2Br \\ | \\ CH^2Br \end{array} \quad + \quad Ag^2O \quad = \quad \begin{array}{c} CH^2 \\ | \quad >O \\ CH^2 \end{array} \quad + \quad 2AgBr$$

　Ethylene bromide.　　Silver oxide.　　Ethylene oxide.　　Silver bromide.

316. Chlorides, Bromides, etc.—The class of simple ethers includes the chlorides, bromides, iodides, sulphides, etc. of the radicals, and in general those compounds which correspond to the simple salts of the metals,—that is, those salts formed by an acid containing no oxygen.

These compounds may be made by heating together the corresponding acid and alcohol. If methyl alcohol is heated with strong hydrochloric acid, a colorless gas, methyl chloride, CH^3Cl, is disengaged.

317. ETHYL IODIDE, C^2H^5I.—In a glass flask we put some alcohol with about two-thirds its weight of amorphous phosphorus; to this flask we fit a cork, through which passes a bottle-shaped tube called an adapter. We have partially closed the lower end of the adapter with some broken glass, and on this have placed a mixture of broken glass with a quantity of iodine equal to two and three-fourths times the weight of the alcohol. The upper end of the adapter is connected with a condenser so inclined that liquid may flow from it into the flask (Fig. 94). When we

FIG. 94.

heat the flask, the alcohol boils, and its vapor, condensing, runs back through the iodine, which is dissolved and brought gradually in contact with the amorphous phosphorus. Phosphorus tri-iodide, PI^3, is then formed; but this immediately reacts with the alcohol, forming ethyl iodide and phosphorous acid. The whole reaction is expressed in the equation

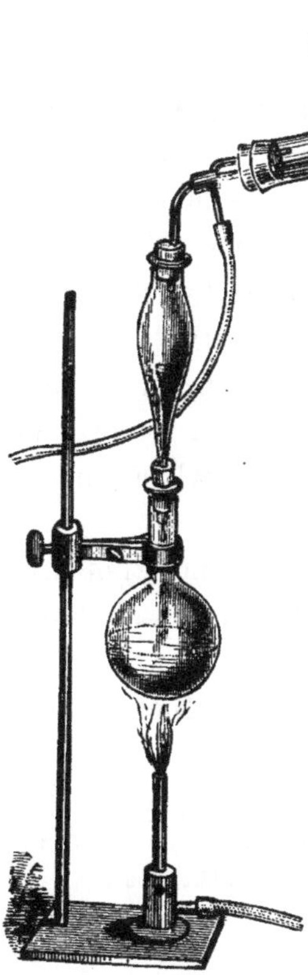

$$3C^2H^5.OH \ + \ P \ + \ I^3 \ = \ 3C^2H^5I \ + \ P(OH)^3$$
Alcohol.   Ethyl iodide. Phosphorous acid.

When the reaction has terminated, we agitate the contents of the flask with a dilute solution of sodium hydrate, and decant the aqueous liquid from the heavy oily layer of ethyl iodide. We then remove all traces of water from the latter by shaking it with some fragments of calcium chloride, and may further purify it by fractional distillation.

Ethyl iodide is a colorless liquid, having at 0° a density of 1.975. It boils at 72°.

318. ETHYL BROMIDE, C^2H^5Br, may be made by distilling a mixture of alcohol, potassium bromide, and sulphuric acid diluted with its weight of water, the substances being used in the proportions required by the equation

$$C^2H^5.OH \ + \ KBr \ + \ H^2SO^4 \ = \ C^2H^5Br \ + \ KHSO^4 \ + \ H^2O$$

It is a liquid having a pleasant odor, and boiling at 40°. It is sometimes used as an anæsthetic.

319. By various means the hydrogen atoms of these simple ethers may be replaced by chlorine, bromine, or iodine, and compounds are then formed which we may consider as derived from other radicals. By the replacement

of a hydrogen atom in ethyl bromide by a bromine atom, we may obtain either ethylene bromide, CH^2Br-CH^2Br, or an isomeric compound called ethylidene bromide, CH^3-CHBr^2. One of the more important of the substances derived by the continued replacement of the hydrogen atoms of a simple ether, is chloroform.

320. CHLOROFORM, $CHCl^3$, may be made by passing methyl chloride mixed with chlorine over charcoal heated to about 200°. It is usually manufactured by distilling a mixture of alcohol and chlorinated lime, commonly called bleaching powder. The reaction is very complex. Chloroform and water condense together in the receiver, and the chloroform separates in a heavy oily layer, for it is hardly soluble in water. It is decanted, shaken first with water and then with a solution of potassium carbonate to remove impurities, and then distilled with calcium chloride, which removes the little water which it held in solution.

Chloroform is a colorless liquid having a pleasant, stimulating odor, and a sweet, burning taste. Its density is 1.5, and it boils at 60.8. It is not inflammable, and communicates a green tint to a flame in which a drop of it is introduced on the end of a glass rod. It is used as an anæsthetic.

321. There is a bromoform, $CHBr^3$, a heavy, colorless liquid, and iodoform, CHI^3, a yellow, crystalline solid. They are formed by the action of bromine and iodine on alcohol in presence of alkaline hydrates or carbonates.

LESSON XL.

ALDEHYDES, CARBON ACIDS, AND COMPOUND ETHERS.

322. In a small beaker, we mix some ordinary alcohol with a little potassium dichromate and some strong sulphuric acid. The mixture becomes warm, and the red color of the potassium dichromate is changed to green. Potassium dichromate, a compound derived from chromic acid, contains much oxygen, and it is reduced by the alcohol, which becomes oxidized. The peculiar odor, somewhat resembling that of apples, which is developed in the

beaker glass, is due principally to a substance called aldehyde. Its composition is C^2H^4O, and it represents alcohol in which the hydroxyl and one hydrogen atom are replaced by an atom of oxygen.

$$CH^3\text{-}CH^2.OH + O = CH^3\text{-}CHO + H^2O$$
Alcohol. Aldehyde.

Aldehyde is made by distilling a mixture of alcohol, sulphuric acid, and potassium dichromate, and condensing the product in a receiver surrounded by ice. It is a very volatile liquid, boiling at 21°.

323. To aldehyde correspond compounds of the same nature derived from each primary alcohol. In each of them the hydroxyl and one atom of hydrogen of the corresponding alcohol are replaced by an atom of oxygen.

324. There is an interesting derivative of aldehyde produced by the prolonged action of chlorine on absolute alcohol. It is an oily liquid, called *chloral*, and represents aldehyde in which three hydrogen atoms are replaced by three atoms of chlorine.

$$CCl^3\text{-}CHO \qquad CH^3\text{-}CHO$$
Chloral. Aldehyde.

Chloral, or trichloraldehyde, combines directly with water, forming a crystalline compound called chloral hydrate. This is much used in medicine for its sleep-producing properties.

325. We may consider that an aldehyde is an oxidation product of an alcohol. If the oxidation proceed still further, the aldehyde is in its turn converted into an acid, and the conversion of an alcohol into an acid may take place without the previous formation of an aldehyde.

Indeed, a carbon acid may be considered as a primary alcohol in which the hydroxyl remains, but the two atoms of hydrogen related to the same carbon atom as the hydroxyl are replaced by an atom of oxygen.

$$CH^3\text{-}CH^2.OH + O^2 = CH^3\text{-}CO.\theta H + H^2O$$
Alcohol. Acetic acid.

We consider, then, that a carbon acid is a compound containing the monatomic group of atoms CO.OH, and this group is often called *carboxyl*. As in all acids, the hydrogen of the hydroxyl group may be replaced by metal, and salts are so formed. If there be only one carboxyl group in the acid, there can be only one series of salts; but if there be two carboxyl groups, we can understand that either both or only one of the hydrogen atoms may be replaced, and there will be two series of salts, neutral salts and acid salts.

326. **Formic Acid,** HCO.OH, is the acid formed by the oxidation of methyl alcohol. We have already seen that it may also be produced from hydrocyanic acid. It exists naturally in certain insects, and it takes its name from its existence in ants. It is made by distilling oxalic acid with glycerin, taking care that the temperature of the mixture does not rise above 100°. The oxalic acid then forms with the glycerin a compound which is again decomposed by the heat, and dilute formic acid distils, while the glycerin is regenerated.

$$C^2O^4H^2 \quad = \quad HCO.OH \quad + \quad CO^2$$
$$\text{Oxalic acid.} \qquad \text{Formic acid.}$$

Formic acid is a colorless, very acid liquid, having a pungent odor. It freezes at 8.5°, and boils at 99°. It mixes with water in all proportions. To a little formic acid in a test-tube, we add some strong sulphuric acid, and gently heat the tube: an effervescence takes place, and we may light the escaping gas at the mouth of the tube. It is carbon monoxide, for the formic acid has been decomposed into that gas and water.

$$HCO.OH \quad = \quad CO \quad + \quad H^2O$$

By replacement of the hydrogen of the hydroxyl in formic acid, *formates* are produced: they are soluble in water, and yield carbon monoxide when heated with sulphuric acid.

327. **Acetic Acid,** $C^2H^4O^2 = CH^3-CO.OH$, is obtained in large quantities during the manufacture of charcoal by the distillation of wood in closed vessels (§ 226). The liquids which condense in this operation consist of tarry matter, dilute acetic acid, wood-spirit, and some other substances. After the tar has been separated, the acid liquid is neutralized with lime, and a crude calcium acetate, generally called pyrolignite of lime, is so formed.

This is mixed with sodium sulphate, and the sodium acetate and insoluble calcium sulphate formed are separated by filtration.

$$Ca(C^2H^3O^2)^2 \quad + \quad Na^2SO^4 \quad = \quad CaSO^4 \quad + \quad 2NaC^2H^3O^2$$

Calcium acetate.   Sodium acetate.

The sodium acetate is then purified by crystallization, and, by heating it with-strong sulphuric acid, is again converted into sodium sulphate and acetic acid which distils.

328. Vinegar is a dilute acetic acid produced by the oxidation of alcohol. The oxidation is brought about by a minute organized ferment which has the property of absorbing oxygen from the air and transferring it to the alcohol. The change is called the acetic fermentation : it does not take place in strong alcohol. In one method of manufacture, the dilute alcohol, or wine, is allowed to trickle over beech-wood shavings contained in a large cask having a double bottom and numerous perforations for the circulation of air (Fig. 95). A large number of these casks are placed in rows, and the shavings are first saturated with some beet-juice' or sour wine in which the ferment is already developed. The slow oxidation of the alcohol produces so much heat that the temperature rises to 30°. It is

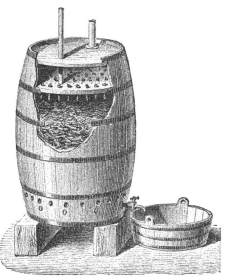

FIG. 95.

usually necessary to allow the same liquid to pass twice through the cask before all of the alcohol is changed to vinegar.

329. Pure acetic acid is a corrosive liquid, having a pungent odor. Its density at 0° is 1.08 ; it freezes at 17°, and boils at 118°. It is soluble in all proportions of water and alcohol.

'In a test-tube we neutralize a few drops of acetic acid with a fragment of solid potassium hydrate : then we introduce a few grains of arsenious oxide, and heat the tube. Dense white vapors, having a very unpleasant garlicky odor, are disengaged. These are due to the formation of a very poisonous compound called cacodyl. The test enables us to recognize an acetate.

330. **Acetates.**—Acetic acid contains only one atom of hydrogen capable of replacement by metal, and the acetates must contain one atom of a metal united with one or more groups, $C^2H^3O^2$, according to the atomicity of the metal.

331. Sodium Acetate, $NaC^2H^3O^2$, crystallizes in large colorless prisms containing three molecules of water. The crystals effloresce in dry air, and the water may be entirely driven out by heat. It is very soluble in water.

332. Lead Acetate, $Pb(C^2H^3O^2)^2$, is commonly called sugar of lead. Its crystals also contain three molecules of water, and have a sweet taste. It is made by dissolving lead oxide, PbO, in acetic acid. Solutions of lead acetate are capable of dissolving an excess of lead oxide, and when carbon dioxide is passed through the liquid, lead carbonate is precipitated, while the neutral acetate remains in solution. Lead acetate is poisonous.

333. Copper Acetate, $Cu(C^2H^3O^2)^2 + H^2O$, forms beautiful bluish-green crystals. *Verdigris* is a combination of copper acetate and cupric oxide, CuO.

334. Before leaving acetic acid, we must study one interesting manner of its formation. We know that hydrochloric acid or an alkaline hydrate will convert hydrocyanic acid into formic acid or an alkaline formate ($\S$ 263). If the hydrogen of hydrocyanic acid be replaced by a methyl group, CH^3, methyl cyanide is obtained, $CH^3.CN$: when this methyl cyanide is boiled with potassium hydrate, ammonia is disengaged, and the solution contains potassium acetate.

$$CH^3.CN \quad + \quad KOH \; + \; H^2O \; = \; NH^3 \quad + \quad CH^3.CO.OK$$
Methyl cyanide.   Potassium acetate.

Since we have found such strong reasons for believing that the two carbon atoms of acetic acid are combined together, we must believe that the two carbon atoms in methyl cyanide are so combined; in hydrocyanic acid the hydrogen atom which occupies the same relation as the carbon of the methyl group in methyl cyanide must then also be combined directly with the carbon.

This conversion of methyl cyanide into acetic acid is only an example of a general reaction of the cyanides of carbon radicals; by boiling with potassium hydrate, the nitrogen atom is always changed for an atom of oxygen and the group OK, thus forming a salt of that carbon acid which contains one more carbon atom than the radical of the cyanide.

335. The general formula of the series of acids derived from the monatomic alcohols is $C^nH^{2n}O^2$. The higher members of this series form part of the natural fats, and the series is generally called the series of fatty acids. The third acid is propionic acid, $C^3H^6O^2$. There are two butyric acids : one of them exists in butter, and the other may be obtained by the action of potassium hydrate on isopropyl cyanide. There are three valeric acids, $C^5H^{10}O^2$; the most com-

mon exists naturally in valerian root and angelica root. It is a colorless liquid, having a strong and unpleasant odor.

336. **Compound Ethers.**—In a glass flask connected with a good condenser (Fig. 96) we distil a mixture of strong alcohol

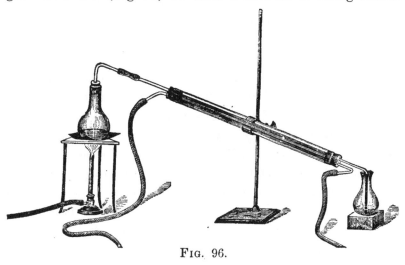

FIG. 96.

with nearly twice its weight of strong sulphuric acid and three times its weight of crystallized sodium acetate. A colorless, volatile liquid, having a fragrant odor, condenses in the receiver. This body is ethyl acetate, and has been formed by the replacement of the sodium atom in sodium acetate by an ethyl group.

$$NaC_2H_3O_2 + C_2H_5.OH + H_2SO_4 = C_2H_5.C_2H_3O_2 + NaHSO_4 + H_2O$$

Sodium acetate.   Ethyl acetate.

Ethyl acetate is a compound ether; the compound ethers are formed by the replacement of the basic hydrogen in any oxygen acid by a hydrocarbon radical. They are the salts of the hydrocarbon radicals, just as sodium acetate is a salt of sodium. Their corresponding acids may be carbon acids or some of the other acids which we have already studied: they may be formed and decomposed by double decomposition, like the salts of the metals. We mix a little ethyl iodide with an alcoholic solution of silver nitrate; a yellow precipitate of silver iodide is formed, and the liquid contains ethyl nitrate.

$$C_2H_5I + AgNO_3 = Ag I + C_2H_5.NO_3$$

Ethyl iodide.   Silver nitrate.   Silver iodide.   Ethyl nitrate.

We gently heat some ethyl acetate with an alcoholic solution of potassium hydrate; the odor of the ethyl acetate disappears; potassium acetate and alcohol have been formed.

$$C^2H^5.C^2H^3O^2 \ + \ KOH \ = \ KC^2H^3O^2 \ + \ C^2H^5.OH$$

We see by these reactions that the compound ethers are precisely analogous to the metallic salts.

337. These ethers are odorous substances, and many of those in which both basic and acid radical are carbon compounds, exist naturally in fruits, of which the odors are due to the ethers. Ethyl formate, $C^2H^5.CHO^2$, exists in rum; ethyl valerate and amyl acetate probably exist in pineapples and bananas. These ethers may be prepared artificially by processes analogous to that described for ethyl acetate, or they may be made by passing hydrochloric acid gas through a mixture of the corresponding acid and alcohol. In this case water and a simple ether (chloride) are first formed, and the latter at once reacts with the carbon acid, the result being a compound ether, while hydrochloric acid is regenerated.

LESSON XLI.

CARBON ACIDS AND COMPOUND ETHERS (2).

338. **Fatty Acids.**—As the number of carbon atoms in the fatty acids increases, these substances are more oily in nature and more insoluble in water. They are liquids at ordinary temperatures until the molecule contains nine atoms of carbon, $C^9H^{18}O^2$; the others are solids, and the melting point is higher as the composition is more complex. Compound ethers of these acids exist in various vegetable products and in animal secretions. The peculiar odors of animals are due to fatty acid ethers. We must pass by the intermediate members of the series and study more particularly those which are most largely used in the arts.

339. PALMITIC ACID, $C^{16}H^{32}O^2$, exists in palm oil, where the radical of the acid is combined with the radical C^3H^5 of glycerin.

It is manufactured by distilling palm oil in a current of super-heated steam : glycerin and palmitic acid are formed, and the latter solidifies to a white mass on cooling. This mass is strongly pressed, to remove a liquid acid, oleic acid, which, existing also in a glycerin compound in the palm oil, is formed at the same time. The palmitic acid is then used for the manufacture of soap and candles.

340. MARGARIC ACID, $C^{17}H^{34}O^2$, exists in nearly all solid fats, and in olive oil. It forms white, crystalline scales, fusible at $60°$.

341. STEARIC ACID, $C^{18}H^{36}O^2$, forms a large proportion of tallow, and may be made by decomposing that substance by super-heated steam. It is a white solid, fusible at $69°$. It dissolves in alcohol and ether, and may be crystallized from its solutions. With the exception of the stearates of potassium, sodium, and ammonium, the salts of stearic acid are insoluble in water.

342. **Oleic Acid,** $C^{18}H^{34}O^2$.—Olive oil contains the glycerin compound of an acid which does not belong to the series of fatty acids. It is an unsaturated carbon compound, and its molecule contains two atoms of hydrogen less than that of stearic acid. It is called oleic acid, and exists in many oils and fats, but always mixed with certain of the fatty acid compounds. It is an oily liquid, which freezes at $4°$.

343. **Fats and Oils.**—The natural fats and fatty oils are compound ethers in which a glycerin radical replaces the basic hydrogen of the fatty acids. We have already seen that glycerin is a triatomic alcohol: it contains three hydroxyl groups, and in the fats and oils each hydroxyl group is replaced by a fatty acid less the hydrogen of its hydroxyl. The natural fats must, then, represent three molecules of fatty acid and a molecule of glycerin. The names of these fatty bodies are derived from those of the fatty acids which take part in their formation.

344. PALMITIN, $C^3H^5(C^{16}H^{31}O^2)^3$, may be extracted from palm oil which has been solidified by cold and then subjected to pressure to remove the liquid fatty matters. It is a white solid, melting at $60°$.

345. MARGARIN, $C^3H^5(C^{17}H^{33}O^2)^3$, and STEARIN, C^3H^5-

$(C^{18}H^{35}O^2)^3$, are also solids; they exist in the solid fats, such as tallow.

346. OLEIN, $C^3H^5(C^{18}H^{33}O^2)^3$, constitutes the greater portion of olive oil, almond oil, and other analogous oils. It is a liquid, which solidifies at 10°.

Oils are usually classed as *fat oils* and *drying oils*. The first are such as do not solidify on exposure to air, but become rancid and acquire an unpleasant odor. They are numerous, and include olive oil, cotton-seed oil, oil of sweet almonds, peanut oil, and many others. The drying oils, of which the type is linseed oil, absorb oxygen and become thick and hard when exposed to the air; they are used in the preparation of paints and varnishes.

347. **Saponification.**—The decomposition of a compound ether by a metallic hydrate, a decomposition which results in the formation of a metallic salt and an alcohol, is in chemical language called saponification; however, a more restricted sense of the word implies the decomposition of a fatty body, with the formation of soap and glycerin. We boil some palm oil or olive oil with a solution of sodium hydrate; the oil disappears, and a soap has been formed, while glycerin is set free in the liquid. It is necessary that ordinary soaps shall be soluble in water, and we have already seen that the only ordinary metals which yield soluble salts with the fatty acids are potassium and sodium; in other words, the alkaline metals (§ 341). Soap, then, is an alkaline salt of one of the higher fatty acids, generally palmitic, stearic, and margaric acids, to which must be added oleic acid. Soft soaps are made with potassium hydrate, while sodium hydrate yields the hard soaps.

In the manufacture of soap, the fat or oil is first boiled with a rather weak solution of sodium hydrate, generally known as concentrated lye, and, when the mixture becomes pasty, enough strong sodium hydrate is added to saponify the fat completely. To separate the excess of water, common salt is added; this dissolves in the water, causing the soap to come to the surface, for common soap is insoluble in salt water. The salty water, containing the excess of alkaline hydrate employed, is then drawn off, and the soap hardens on cooling. As it is not easy to separate from the waste liquid the glycerin formed in the reaction, it is more economical to decompose the fat by superheated steam, and boil with sodium hydrate the fatty acid which floats on the dilute glycerin. While soap is soluble in water, it is decomposed by a large quantity of that liquid, a small quantity of alkaline hydrate being set free, while the fatty acid becomes insoluble. The free alkali produces the cleansing effects, and the fatty acid forms the lather: we know that soap will not produce a lather if we use too little water. Ordinary soap is insoluble in salt water, but a soap which is soluble in salt water may be made from cocoanut oil; it is called saltwater soap. It contains an alkaline laurate and myristate, *lauric acid*, $C^{12}H^{24}O^2$, and *myristic acid*, $C^{14}H^{28}O^2$, existing as glycerin ethers in cocoanut oil.

348. STEARIN CANDLES are made from a mixture of solid fatty acids obtained by saponifying tallow by superheated steam and a small quantity of lime. The small quantity of insoluble calcium soap so formed is decomposed by sulphuric acid, and the oleic acid is separated from the solid acids by pressing the mass between warm plates. The oleic acid is used for the manufacture of soap. Certain fatty bodies, among them palm oil, are entirely decomposed by superheated steam, without the aid of lime. In this reaction the water acts as would either an acid or an alkaline hydrate, part of its molecule completing the basic molecule of glycerin, while the other part completes the acid molecule.

$$C^3H^5(C^{16}H^{31}O^2)^3 \quad + \quad 3HOH \quad = \quad C^3H^5(OH)^3 \quad + \quad 3C^{16}H^{32}O^2$$
Palmitin. Glycerin. Palmitic acid.

The saponification of fats and oils may be brought about by the action of strong acids, such as sulphuric acid, for the strong acid forms a new compound ether with the glycerin radical, and sets the fatty acid free; the compound ether may then be again decomposed into glycerin and acid by the addition of water.

LESSON XLII.

CARBON ACIDS (3).

349. **Lactic Acid,** $C^3H^6O^3$, is a product of the fermentation of milk, and of a peculiar fermentation which glucose undergoes in the presence of basic substances which will neutralize the acid as fast as it is formed. It is usually made by allowing a solution of glucose to which some sour milk, a little old cheese, and some chalk have been added, to ferment in a warm place until the whole is converted into a solid mass of calcium lactate. This salt is then purified by crystallization, and is decomposed by the exact quantity of sulphuric acid required to precipitate all of the calcium in the form of insoluble calcium sulphate. The solution of lactic acid is then separated by a filter, and evaporated on a water-bath. Lactic acid remains as a colorless, very sour, syrupy liquid, which is decomposed when heated.

Lactic acid is propionic acid, $C^3H^6O^2$, in which one atom of hydrogen is replaced by a hydroxyl group; it is consequently at the same time an alcohol and an acid. Chemists have obtained another acid of the same composition, an isomeride of lactic acid, and the differences of the two are due to different positions of the hydroxyl group. The isomeric lactic acid is called *hydracrylic acid,* because it is decomposed by heat into water and an acid called acrylic acid, $C^3H^4O^2$.

CH^3–CH^2–CO.OH	CH^3–CH(OH)–CO.OH	CH^2(OH)–CH^2–CO.OH
Propionic acid.	Lactic acid.	Hydracrylic acid.

350. **Oxalic Acid,** $C^2H^2O^4$, exists naturally in many plants; it gives the sour taste to sour grass, and at certain seasons is present in small quantities in rhubarb-leaves. It is a product of the oxidation of many vegetable matters: it may be made by boiling starch with rather dilute nitric acid, and evaporating the liquid. It is now manufactured by heating to 200° a pasty mixture of saw-dust and potassium hydrate; potassium oxalate is so formed, and is separated by treating the mass with hot water, in which it is quite soluble. The solution of potassium oxalate is

then mixed with milk of lime, which is calcium hydrate, and in-soluble calcium oxalate is formed, while the solution contains potassium hydrate, which is used for another operation.

$$K^2C^2O^4 \quad + \quad Ca(OH)^2 \quad = \quad CaC^2O^4 \quad + \quad 2KOH$$

Potassium oxalate.   Calcium hydrate.   Calcium oxalate.

The calcium oxalate is decomposed by sulphuric acid, which forms insoluble calcium sulphate, and the solution of oxalic acid is evaporated until it is strong enough to crystallize.

Oxalic acid forms large, colorless prisms, containing two mole-cules of water of crystallization. In dry air, these crystals efflo-resce, and the anhydrous acid may be obtained by carefully heating them to 100°. Oxalic acid dissolves in fifteen times its weight of cold water, and is also soluble in alcohol. When heated to about 150°, it is decomposed with formation of carbon monoxide, carbon dioxide, formic acid, and water.

$$2C^2H^2O^4 \quad = \quad CO \quad + \quad 2CO^2 \quad + \quad CH^2O^2 \quad + \quad H^2O$$

Oxalic acid.   Formic acid.

We have already learned that both carbon monoxide and formic acid are prepared by the decomposition of oxalic acid.

351. We neutralize a solution of oxalic acid by the addition of a little ammonia-water, and then pour into it some solution of calcium chloride. A white precipitate of insoluble calcium oxalate is formed.

$$(NH^4)^2C^2O^4 \quad + \quad CaCl^2 \quad = \quad 2NH^4Cl \quad + \quad CaC^2O^4$$

Ammonium oxalate.   Calcium chloride.   Ammonium chloride.   Calcium oxalate.

Oxalic acid is poisonous; its antidote is chalk, which is calcium carbonate: this causes the formation of insoluble calcium oxalate.

We have prepared some silver oxalate by adding solution of silver nitrate to a solution of oxalic acid neutralized with ammonia. The insoluble silver oxalate is separated by filtration and dried. When we heat a small quantity of this powder in a test-tube, it suddenly explodes, being decomposed into carbon monoxide, carbon dioxide, and silver.

Since oxalic acid contains two carboxyl groups, CO.OH, it is a dibasic acid; it contains two atoms of replaceable hydrogen, and with monatomic metals may form two series of salts, acid oxalates, in which only one atom of hydrogen is replaced by metal, and neutral salts, in which both atoms are so replaced.

One atom of a diatomic metal like calcium will of course replace both hydrogen atoms.

With the exception of the oxalates of potassium, sodium, and ammonium, all of the neutral oxalates of the ordinary metals are insoluble in water, but dissolve in dilute acids such as sulphuric and hydrochloric acids.

352. Tartaric Acid, $C^4H^6O^6$, is the acid of grapes. In the casks in which wine is kept, there is deposited an impure potassium acid tartrate, called *argol*. This is purified by crystallization from boiling water, and the pure potassium acid tartrate so obtained constitutes cream of tartar. By boiling the cream of tartar with chalk, and adding · sufficient calcium chloride to form potassium chloride with the potassium, insoluble calcium tartrate is formed, while carbon dioxide is given off, and potassium chloride remains in solution. The calcium tartrate is separated by filtration, and, after being washed with water, is decomposed by the exact quantity of dilute sulphuric acid. Calcium sulphate is precipitated, and when the filtered solution has been sufficiently concentrated by evaporation, crystals of tartaric acid are formed.

Tartaric acid is in large, prismatic crystals, soluble in about half their weight of cold water, and also soluble in alcohol. By the action of heat it is converted into several other acids, of which the compositions depend on the temperature at which the tartaric acid is decomposed.

353. We can easily understand the molecular constitution of tartaric acid by studying that of substances to which it is intimately related. When ethylene cyanide $(CN)CH^2-CH^2(CN)$ is boiled with potassium hydrate, ammonia is disengaged, and there is formed the potassium salt of succinic acid, so called because it is formed by the action of heat on amber.

$$
\begin{array}{ll}
CH^2-CN & \\
\quad | & + \ 4H^2O \ = \\
CH^2-CN &
\end{array}
\quad
\begin{array}{l}
CH^2-CO.OH \\
\quad | \qquad\qquad + \ 2NH^3 \\
CH^2-CO.OH
\end{array}
$$

Ethylene cyanide.   Succinic acid.

There exists in apples, gooseberries, and many other fruits an acid called malic acid, and this has also been prepared artificially in such a manner as to show that it represents succinic acid in which one atom of hydrogen is replaced by a hydroxyl group. The replacement of two hydrogen atoms of succinic acid by hydroxyl groups yields tartaric acid.

$$
\begin{array}{lll}
CH^2-CO.OH & CH(OH)-CO.OH & CH(OH)-CO.OH \\
\quad | & \quad | & \quad | \\
CH^2-CO.OH & CH^2-CO.OH & CH(OH)-CO.OH
\end{array}
$$

  Succinic acid.   Malic acid.   Tartaric acid.

Tartaric acid is, then, a diatomic alcohol, for it contains two hydroxyl groups related to two carbon atoms, and it is a dibasic acid, for it contains two carboxyl groups, CO.OH.

Of monatomic metals like potassium and sodium there are two series of tartrates, acid tartrates, in which only one atom of basic hydrogen is replaced, and neutral tartrates, in which both are replaced.

354. POTASSIUM ACID TARTRATE, $KC^4H^5O^6$, is cream of tartar, and is made by simply purifying the argol of wine-casks. It is almost insoluble in cold water, but dissolves in boiling water.

When cream of tartar is heated to redness, it leaves a residue of charcoal and pure potassium carbonate, which may be dissolved from the mass by water. Pure potassium carbonate is usually obtained in this manner.

355. POTASSIUM TARTRATE, $K^2C^4H^4O^6$, is made by adding potassium carbonate to a boiling solution of cream of tartar as long as carbon dioxide is disengaged. When the concentrated solution cools, the salt separates in crystals which are very soluble in water.

356. POTASSIUM SODIUM TARTRATE, $KNaC^4H^4O^6$, is commonly called Rochelle salt. It is made by neutralizing with sodium carbonate a boiling solution of cream of tartar. It forms beautiful, colorless crystals, freely soluble in water.

357. ANTIMONIO-POTASSIUM TARTRATE, $K(SbO)C^4H^4O^6$, known as tartar emetic, is formed when antimonous oxide is boiled with cream of tartar. Its crystals contain one molecule of water of crystallization for every two molecules of the salt, and effloresce in dry air. It is soluble in water, and is poisonous. When hydrogen sulphide is passed through its solution, an orange-colored precipitate of antimony sulphide is formed.

358. **Citric Acid,** $C^6H^8O^7$, exists in lemons, oranges, currants, and many other fruits. It is made by allowing the juice of lemons or sour oranges to stand until it begins to ferment, and then neutralizing the boiling filtered liquid with chalk. The insoluble calcium citrate formed is washed with boiling water, and decomposed by dilute sulphuric acid; citric acid crystallizes from the solution separated from the insoluble calcium sulphate.

Citric acid forms large colorless crystals, soluble in about three-fourths their weight of cold water, and having a very sour taste. It is a tribasic acid, containing three carboxyl groups. Its cold solutions are not precipitated by lime-water, but become turbid when the liquid is boiled, for calcium citrate is more soluble in cold than in hot water. Magnesium citrate is employed as a purgative in medicine.

LESSON XLIII.

HYDRATES OF CARBON,

359. Plants and vegetables contain a number of compounds composed of carbon, hydrogen, and oxygen, the last two elements being present in exactly the proportions required for the formation of water. For that reason these compounds are called carbohydrates, or hydrates of carbon. The compositions of these substances are expressed by three different formulæ, but there are a number of isomeric compounds for each formula. Of all these substances we can study only the more common, one or two of each series, and the types of the three series are

$$C^6H^{10}O^5 \qquad C^6H^{12}O^6 \qquad C^{12}H^{22}O^{11}$$
$$\text{Starch.} \qquad \text{Glucose.} \qquad \text{Saccharose.}$$

360. **Starch** is found everywhere in the vegetable kingdom, and constitutes the greater part of all grains and of many tuberose roots like the potato. It is obtained by reducing potatoes to a pulp, and washing this pulp in a sieve through which flows a stream of water. The fibrous matters, consisting of the torn cells of the potato, remain in the sieve, while the small particles of starch pass through and are deposited from the water, which is allowed to flow slowly down long inclined planes. From grains the starch is extracted by grinding the grain to flour, and kneading the flour in a sieve under running water. The starch passes through, as before, while the nitrogenized matter of the grain forms a soft, elastic mass, called *gluten*.

The starch so obtained is simply separated from the vegetable cells in which it was formed. It occurs as a fine powder, in which microscopic examination reveals a peculiar granular structure. The size and shape of these granules vary with the source of the starch (Fig. 97): they are from 2 to 185 thousandths of a millimetre in diameter. They are formed of concentric layers, and their structure becomes apparent when a little starch is dried at 100°, and, after moistening with a drop of water containing a trace of iodine, is examined by the aid of a microscope. The granules

Fig. 97.   Fig. 98.

then swell, and, as the exterior layers burst, the interior structure is exposed (Fig. 98).

Starch is insoluble in water and alcohol; but, when it is rubbed with water in a mortar with rough sides, a small quantity of the interior of the granules appears to dissolve. When it is boiled with a large quantity of water, the granules burst, and a turbid liquid is obtained on cooling; this contains some soluble starch, and holds in suspension the insoluble starch. When heated with water to 60° or 70°, starch forms a gelatinous mass, called starch paste. We have already seen that starch develops a blue color with iodine; and as starch is the test for iodine, so iodine is the test for starch. It is probable that the blue substance is only starch dyed by iodine; for, if dried and exposed to the air, its color gradually fades as the iodine volatilizes.

While the composition of starch is represented by the formula $C^6H^{10}O^5$, it is probable that this formula does not express a molecule of starch, but that the molecule contains a multiple of $C^6H^{10}O^5$.

Boiling with dilute acids converts starch into glucose.

$$C^6H^{10}O^5 \quad + \quad H^2O \quad = \quad C^6H^{12}O^6$$

By the influence of a peculiar substance called diastase, which is formed during the germination of grain, starch is converted into an isomeride of glucose, to which the name maltose has been given.

361. Dextrin.—When starch is heated to between 160° and 200°, it is changed into a body which is soluble in water, and which is not colored by iodine. It is a pale-yellow powder, called dextrin. · Its solution is gummy, and is used as a mucilage.

362. Cellulose. contains the same proportions of carbon, hydrogen, and oxygen as starch. It is the matter which forms the walls of young cells in vegetables, and is deposited, together with other matters, in the older cells. Linen, cotton, paper, and the pith of certain plants are almost pure cellulose, which may be obtained by washing linen or cotton successively with dilute solution of potassium hydrate, water, chlorine-water, acetic acid, alcohol, ether, and water. The insoluble matter left after these operations is cellulose.

It is a translucid, white solid, having a density of about 1.3. It is not soluble in either water, alcohol, dilute acids, or alkaline hydrates. It dissolves, however, in the blue liquid obtained by shaking copper with ammonia-water in contact with the air. By the action of strong sulphuric acid on cellulose, a gummy mass is obtained, which long boiling with water converts into fermentible glucose.

When paper is soaked in a cold mixture of sulphuric acid with half its volume of water, and is then thoroughly washed and dried, it is converted into a semi-transparent substance, which is called vegetable parchment. This is the substance generally used for dialysis (§ 220).

363. GUN-COTTON is made by soaking cotton wool in a mixture of about equal volumes of strong nitric and sulphuric acids, and washing the product in running water until the last traces of acid are removed. After drying in the air, the substance has all the appearances of cotton, but is not as soft to the touch. It is very inflammable, and burns with a flash, leaving no residue.

In gun-cotton, part of the hydrogen of the cellulose has been replaced by monatomic nitryl groups, NO^2, and the properties of the gun-cotton are modified according to the number of hydrogen atoms so replaced. The most explosive variety is called pyroxylin, and is probably a mixture of dinitrocellulose and trinitrocellulose.

$$C^6H^{10}O^5 \qquad C^6H^8(NO^2)^2O^5 \qquad C^6H^7(NO^2)^3O^5$$

Cellulose.   Dinitrocellulose.   Trinitrocellulose.

A large volume of gas is produced by the explosion, and attempts have been made to substitute this variety of gun-cotton for gunpowder. Gun-cotton is insoluble in water, alcohol, and ether; by careful operations a variety may be obtained which is quite soluble in a mixture of alcohol and ether, and the solution is employed, under the name collodion, in photography and in surgery. The soluble variety is probably dinitrocellulose.

364. **Glucose,** $C^6H^{12}O^6$, exists in many fruits, and when present in large quantity forms a white efflorescence on the surface when the fruits are dried, as is the case with figs and raisins. The solid matter which deposits in old honey is glucose.

Glucose is manufactured by boiling starch with a large quantity of water containing about one-half per cent. of sulphuric acid. The starch is not added until the liquid is boiling, and after about half an hour's cooking it is completely converted into glucose. The sulphuric acid is then neutralized with chalk, and after the insoluble calcium sulphate has been separated by filtration, the solution of glucose is concentrated until it will solidify to a crystalline mass on cooling.

Glucose forms small, rounded, crystalline masses, which contain one molecule of water of crystallization for each molecule of glucose. When cautiously heated, it melts, and again becomes solid at 100°, all the water of crystallization being then expelled. Glucose dissolves in about its own weight of cold water, and the solution has a sweet taste. It is much employed in confectionery and syrups, but to produce the same sweetness in a given quantity of solution, the glucose must be employed in three times the quantity of ordinary sugar which would be required.

In a test-tube we boil a mixture of sodium hydrate solution, potassium and sodium tartrate, and cupric sulphate: this is called a cupro-alkaline solution, and is not changed by heat, but when we add a little glucose to the boiling liquid, the color changes to yellowish red, and on standing red cuprous oxide is deposited. The glucose has reduced the cupric solution: glucose then acts as a reducing agent. The reduction may even result in the separation of metal. To a solution of silver nitrate we add ammonia-water until the precipitate at first formed is just redissolved. Now on adding a little glucose and gently warming the tube a brilliant mirror of silver is formed on its walls. In these reactions the glucose is oxidized and converted into complex acids.

We already know that glucose and maltose are by fermentation decomposed into carbon dioxide and alcohol.

365. **Saccharose, or Cane-Sugar,** $C^{12}H^{22}O^{11}$.—This compound, which is ordinary sugar, is extracted principally from sugar-cane, sugar-maple, beet-root, and sorghum. Maple-sugar flows from incisions made in the bark of the maple. Sugar-cane, beet-root, or the plants from which sugar is to be extracted, are finely cut, and subjected to strong pressure, by which the juice is expressed. The liquid is then heated by steam in large boilers, and milk of lime (calcium hydrate) is added to neutralize the natural acids of the juice and form insoluble compounds with certain nitrogenized principles which are present. The syrup dissolves a considerable quantity of lime, and this is precipitated either by a current of carbon dioxide, or by ammonium phosphate, which forms insoluble calcium phosphate, while ammonia is disengaged. The syrup is then heated, and filtered through a layer of grained animal charcoal, and is afterwards concentrated at as low a temperature as possible by boiling in large vessels in which a vacuum is made by pumps. When it is sufficiently concentrated, the syrup is run into cooling-pans, where it is continually stirred, so that the sugar may separate in small crystals, constituting granulated sugar. This sugar is purified or refined by being again dissolved and fil-

tered through animal charcoal, after which the syrup must be again evaporated and crystallized. Loaf-sugar is made by placing the still wet granulated sugar in conical moulds having a hole at the point, which is the lower portion. The moulds are heated to about 25° in an oven, until the sugar has formed a porous mass, from which the syrup is drained by opening the hole in the mould. A little very strong colorless syrup is then poured in, and as this crystallizes it renders the porous loaf hard and compact. Granulated sugar is dried by rapid rotation in a cylinder of wire gauze, through which the syrup is thrown by centrifugal force; it then constitutes soft sugar. The still moist sugar is dried and converted into dry granulated sugar by being sifted on a revolving cylinder heated by steam and contained in a large, partially open drum.

During the manufacture of sugar a part of that substance is by the action of the heat and water converted into two other substances, glucose, and a body isomeric with it, named levulose.

$$C^{12}H^{22}O^{11} \quad + \quad H^2O \quad = \quad C^6H^{12}O^6 \quad + \quad C^6H^{12}O^6$$

Saccharose.   Glucose.   Levulose.

The mixture of these substances can be crystallized only with great difficulty, and the uncrystallizable syrup constitutes molasses. The purest sugar is rock-candy, and is obtained by stretching threads through a vessel containing a very concentrated syrup. The sugar then deposits in large crystals on the threads.

Sugar is insoluble in ether and in absolute alcohol. Its crystals are anhydrous. It melts at 160°, and on cooling forms a hard, amorphous mass. At about 190° it is partially decomposed, yielding a brown, bitter substance known as caramel. It does not reduce cupro-alkaline solutions, but by long boiling is converted into glucose, which then effects the reduction.

366. **Lactose** is a hard, not very sweet substance which exists in the milk of animals, and is usually made by simply evaporating the whey left in the manufacture of cheese. It has the same composition as saccharose, but its crystals contain one molecule of water of crystallization to one of lactose, $C^{12}H^{22}O^{11} + H^2O$.

367. The saccharine substances, of which we have considered only a few,

p

rotate the plane of polarization of polarized light passed through their solutions.   Glucose, saccharose, and lactose turn it to the right; levulose rotates
it to the left.

368. The gums and mucilages which are obtained from certain plants are
analogous to saccharose in composition.   Gum-arabic, which flows naturally
from several species of acacia, is said to contain a body, *arabin,* having the
composition $C^{12}H^{22}O^{11}$.

LESSON XLIV.

BENZOL DERIVATIVES (1).

369. We have already seen that the unsaturated hydrocarbon
benzol, C^6H^6, acts precisely like the saturated hydrocarbons, in that
its compounds are formed by the replacement of its hydrogen
atoms by other atoms or groups.   We have learned that mono-
chlorobenzol, C^6H^5Cl, is formed in this manner by the replacement
of an atom of hydrogen by one of chlorine.   Since we may consider that the alcohols are formed by the replacement of one or
more hydrogen atoms in the saturated hydrocarbons by the same
number of hydroxyl groups, we can understand that a similar
replacement in benzol should yield substances analogous to the alcohols.   While these substances do resemble the alcohols in many
of their chemical relations, they have at the same time certain
other properties; the hydrogen of their hydroxyl is more readily
replaced by atoms of metal than is that of the alcohols.   They
are called *phenols ;* the most simple is that in which only one
hydrogen atom is replaced by hydroxyl, and it is ordinary phenol,
commonly called carbolic acid.

370. **Phenol, $C^6H^5.OH$.**—This important compound can be
prepared artificially from benzol, but it is always obtained from
coal-tar, for it is one of the products of the destructive distillation
of coal.   After the benzol has been separated from the tar, that
portion which distils during the fractional distillation between
150° and 200° is collected separately, and is mixed with a saturated solution of sodium hydrate.   A compound in which the

hydrogen of the hydroxyl in phenol is replaced by sodium is so formed; this is dissolved in boiling water, and the solution separated from the oily matters, which remain unaffected. The solution of sodium phenate is then treated with hydrochloric acid, the reaction yielding phenol and sodium chloride.

$$C^6H^5.ONa + HCl = C^6H^5.OH + NaCl$$
Sodium phenate.　　　　　　　Phenol.

The phenol is not very soluble in water, and when it has separated is dried with calcium chloride, and distilled. The product is then cooled in a mixture of ice and salt, and the phenol forms crystals which are separated and drained.

Phenol crystallizes in colorless needles, fusible at 35°; it boils at 186°. It has a peculiar characteristic odor, and a burning taste. It is only slightly soluble in water. Although it does not redden blue litmus, it readily reacts with the metallic hydrates, forming crystallizable compounds which in some respects resemble the salts. It acquires a more or less intense red color on exposure to air and light. Phenol is an exceedingly valuable agent for the destruction of low forms of life. It prevents putrefaction and decay of animal and vegetable matters, because it prevents the development of the minute germs of life which are the cause of such decompositions. Phenol is poisonous, and in a concentrated form is quite corrosive to living animal tissues.

When bromine-water is added to even a very dilute solution of phenol, a yellow precipitate of tribromophenol, $C^6H^2Br^3(OH)$, is formed. A pine shaving dipped in phenol and then exposed to the air acquires a blue color. These properties aid us in identifying phenol.

When two hydrogen atoms of benzol are replaced by hydroxyl groups, diatomic phenols, usually called oxyphenols, result. They naturally have the composition $C^6H^4(OH)^2$. Three oxyphenols are known, and we have already seen that we can understand these cases of isomerism by attributing to the hydroxyl groups different positions in the system of carbon atoms which are so intimately related together.

371. **Trinitrophenol,** $C^6H^2(NO^2)^3.OH$, commonly called picric acid, is obtained by boiling phenol with concentrated nitric acid.

$$C^6H^5.OH + 3HNO^3 = C^6H^2(NO^2)^3.OH + 3H^2O$$

It crystallizes from its solution in boiling water in small, lemon-yellow scales, which are not very soluble in cold water. It has an exceedingly bitter taste. It has acid properties, for the three nitryl groups, NO^2, seem to make the hydrogen of the hydroxyl more readily replaceable by metal.

372. POTASSIUM PICRATE, $C^6H^2(NO^2)^3.OK$, may be made by adding potassium carbonate to a boiling solution of picric acid as long as carbon dioxide is disengaged. It forms long yellow needles, only slightly soluble in cold water.

373. AMMONIUM PICRATE, $C^6H^2(NO^2)^3.ONH^4$, is obtained in a similar manner by neutralizing picric acid with ammonia-water. It burns with a flash, without leaving a residue, and has been used in the manufacture of certain kinds of gunpowder and colored fires (§452).

374. **Nitrobenzol,** $C^6H^5.NO^2$.—When benzol is added in small portions to a cold mixture of strong nitric and sulphuric acids, and the liquid is constantly stirred, the benzol dissolves ; when the solution is poured into cold water, a heavy, colorless oil separates. This is nitrobenzol : a hydrogen atom of benzol has been replaced by a nitryl group, NO^2.

$$C^6H^6 \ + \ HNO^3 \ = \ C^6H^5.NO^2 \ + \ H^2O$$

Nitrobenzol freezes at 3°, and boils at 205°. It has an odor resembling that of bitter almonds, and is used in perfumery, especially for imparting an odor to soap. It is manufactured in large quantities for the production of aniline.

375. **Aniline,** $C^6H^5.NH^2$.—If nitrobenzol be treated with a mixture capable of generating hydrogen, the nitryl group is reduced, and converted into a group NH^2. Almost all reducing agents produce this change, but in the arts a mixture of iron filings and acetic acid is used. The hydrogen eliminated from the acetic acid by the iron, with formation of iron acetate, then reduces the nitrobenzol to aniline.

$$C^6H^5.NO^2 \ + \ 3H^2 \ = \ C^6H^5.NH^2 \ + \ 2H^2O$$
$$\text{Nitrobenzol.} \qquad\qquad \text{Aniline.}$$

The operation is conducted in large cylinders in which the nitrobenzol is continually stirred with the acetic acid and iron filings. The aniline formed is then distilled.

Aniline is a colorless liquid, but becomes brown on long ex-

posure to the air. It has an unpleasant odor, and an acrid, burning taste. It is heavier than water, in which it is insoluble. It
boils at 184°. It is soluble in all proportions of alcohol and
ether. It is not alkaline to litmus, but combines directly with
acids, forming crystallizable salts.

Aniline represents ammonia in which one atom of hydrogen is replaced by
the monatomic group C^6H^5. All of the hydrocarbon radicals are capable of
replacing the hydrogen of ammonia, and the compounds so formed are called
compound ammonias, or amines. Thus, methylamine, dimethylamine, and
trimethylamine are formed respectively by the replacement of one, two, and
three atoms of hydrogen in a molecule of ammonia.

NH^3	$NH^2.CH^3$	$NH(CH^3)^2$	$N(CH^3)^3$
Ammonia.	Methylamine.	Dimethylamine.	Trimethylamine.

The group C^6H^5, which is benzol less one atom of hydrogen, is called *phenyl*,
and phenol is then phenyl hydrate, while aniline is phenylamine.

376. To a little aniline in a test-tube, we add a crystal of potassium nitrate, and then some strong sulphuric acid; a bright red
color is produced. In another tube we mix some aniline with
about twice its volume of strong sulphuric acid, and then drop in
a small fragment of potassium dichromate; a magnificent blue
color is developed, and becomes violet when the mixture is diluted
with water. A little bleaching-powder, that is, chlorinated lime,
added to aniline produces also a violet color. These reactions are
applied on a large scale in the manufacture of numerous coloring
matters derived from aniline.

377. **Rosaniline.**—The benzol of commerce is not pure, it contains much methylbenzol or toluol: when it is converted successively into nitrobenzol and aniline, a nitro-derivative of toluol is
also formed, and this is reduced to methylaniline, just as the nitrobenzol is reduced to aniline. When such aniline is heated with
oxidizing agents, both the aniline and the methylaniline lose hydrogen atoms, and the residues of the molecules combine, forming a
complex body called rosaniline.

$$C^6H^7N + 2C^7H^9N + O^3 = C^{20}H^{19}N^3 + 3H^2O$$

Aniline.	Methylaniline.	Rosaniline.	

Large quantities of rosaniline are manufactured by heating
commercial aniline either with arsenic acid or under pressure with

nitrobenzol; the oxygen of the arsenic acid or of the nitrobenzol, removes hydrogen from the aniline.

Rosaniline is a colorless substance, but its salts have magnificent colors and are used as dye-stuffs. The rich red coloring matter called fuchsine is a compound of one molecule of rosaniline with one of hydrochloric acid. If a hot saturated solution of this body be treated with sodium hydrate, the color disappears; sodium chloride is formed, and rosaniline separates as an almost colorless, crystalline precipitate.

The hydrogen atoms of rosaniline may be replaced by various monatomic radicals, such as methyl, ethyl, phenyl. The compounds formed by three such replacements are more easily obtained than the others, and the salts of the resulting tri-substituted rosanilines constitute a numerous, varied, and valuable class of coloring agents, known as the aniline dyes.

LESSON XLV.

BENZOL DERIVATIVES (2).

378. The hydrocarbons derived from benzol by the replacement of its hydrogen atoms by groups such as methyl or ethyl, are capable of forming both phenols and alcohols; for if the replacement of a hydrogen atom by hydroxyl be in the benzol radical, a phenol would result, while an alcohol would be formed by such a replacement in the methyl or ethyl group. Methyl-benzol or toluol can thus form an alcohol, called benzyl alcohol, and three isomeric phenols, which are called cresols.

$$C^6H^5\text{-}CH^3 \qquad HO.C^6H^4.(CH^3) \qquad C^6H^5.CH^2.OH$$

Methyl benzol.   Cresols.   Benzyl alcohol.

Our time will permit the study of only a few of these compounds.

379. **Benzyl Aldehyde,** $C^6H^5.CHO$.—When chlorine gas is

passed through boiling toluol, benzyl chloride, $C^6H^5-CH^2Cl$, is formed, and by alkaline hydrates this may be converted into benzyl alcohol, $C^6H^5-CH^2.OH$. Just as ordinary alcohol may by slow oxidation be converted into aldehyde, benzyl alcohol is by the action of nitric acid converted into benzyl aldehyde. The latter body is interesting, because it is the essential part of oil of bitter almonds, so much used for flavoring. The oil of bitter almonds is, however, poisonous, for it contains hydrocyanic acid, which, together with benzyl aldehyde, results from the action of water on a substance called *amygdalin*, existing in the almonds.

380. **Benzoic Acid**, $C^6H^5.CO.OH$, exists naturally in gum benzoin, and is the product of the oxidation of benzyl aldehyde and benzyl alcohol. It may be easily prepared from gum benzoin, by gently heating some of that resin in a shallow dish, over which is pasted a piece of filter-paper. We cover the dish with a beaker, and the vapor of benzoic acid passes through the paper, on which and in the beaker it condenses in beautiful feathery tufts (Fig. 99).

Benzoic acid crystallizes in colorless needles or thin plates. It melts at 121°, and boils at 250°. It is not very soluble in cold water, but dissolves in about twelve times its weight of boiling water, and is also soluble in alcohol. It is an excellent antiseptic or preservative.

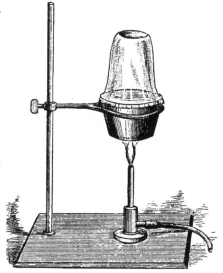

FIG. 99.

381. **Salicyl Aldehyde**, $C^6H^4(OH).CHO$. — The pleasant odor of essential oil of meadow-sweet is due to a compound representing benzyl aldehyde in which an atom of hydrogen in the benzol group is replaced by the radical hydroxyl. It is at the same time an aldehyde and a phenol. It is a colorless liquid,

boiling at 196°. It is heavier than water. Oxidizing agents convert it into

Salicylic Acid, $C^6H^4(OH)CO.OH$, in which the group CHO of the aldehyde has been changed to carboxyl, $CO.OH$. Salicylic acid is now manufactured by the action of carbon dioxide on phenol, or, more correctly, sodium phenate. We may represent the reaction

$$C^6H^5.OH \quad + \quad CO^2 \quad = \quad C^6H^4(OH)CO.OH$$
Phenol.   Salicylic acid.

Salicylic acid occurs as methyl salicylate in oil of wintergreen: the basic hydrogen, that of the hydroxyl group, is here replaced by a methyl group, CH^3.

$$C^6H^4(OH).CO.OH \qquad\qquad C^6H^4(OH).CO.OCH^3$$
Salicylic acid.   Methyl salicylate.

When oil of wintergreen is boiled with potassium hydrate, potassium salicylate and methyl alcohol are formed.

$$C^6H^4(OH)CO.OCH^3 \quad + \quad KOH \quad = \quad C^6H^4(OH)CO.OK \quad + \quad CH^3.OH$$

Salicylic acid crystallizes in needles or prisms which are scarcely soluble in cold water, but very soluble in boiling water, alcohol, and ether. It is largely used as a preservative, and to some extent in medicine.

382 Gallic Acid, $C^6H^2(OH)^3.CO.OH$.—Salicylic acid represents benzoic acid in which one atom of hydrogen is replaced by a hydroxyl group. It is a phenol and an acid. Gall-nuts, which are little excrescences produced by the sting of an insect on the leaves and twigs of certain species of oak, contain a substance which, by continued exposure to air and moisture, undergoes a sort of fermentation. When the liquid is pressed from the dark-colored mass, there remains a compound which may be crystallized from boiling water in long, colorless, silky needles. It is gallic acid, a compound which we may consider as benzoic acid in which three hydrogen atoms are replaced by three hydroxyl groups. It is colorless and odorless. When carefully heated to 100°, it is converted into a white volatile substance known as pyrogallol, or pyrogallic acid, while at the same time carbon dioxide is disengaged.

$$C^6H^2(OH)^3.CO.OH \quad = \quad C^6H^3(OH)^3 \quad + \quad CO^2$$
Gallic acid.   Pyrogallol.

Gallic acid is very soluble in boiling water; not very soluble in cold water. Its solutions, especially if an alkaline hydrate be present, absorb oxygen from the air, and become dark in color. This last property is also common to pyrogallol, a solution of which is used as a reducing agent in photography.

383. Tannin.—The well-known astringent properties of certain plants are due to the presence of compounds known as tannins or tannic acids, of which there appear to be a number of varieties. They possess the property of coagulating albumen and gelatin, and of forming black or nearly black precipitates with salts of iron. Tannin may be extracted from gall-nuts by placing the coarsely-powdered nuts in a funnel and pouring through them ether which is not free from water. As the ether runs through, it retains the coloring matters, while the water in the ether dissolves the tannin, and the aqueous solution separates from the layer of ether on standing. When this solution is evaporated at a gentle heat, the tannin remains as a light, very porous mass. It has an astringent taste, and is very soluble in water. When exposed to moist air, it is converted into gallic acid; a temperature of about 210° decomposes it, with formation of pyrogallol and carbon dioxide. These reactions indicate that tannin is related to gallic acid, and at least one of its varieties appears to be formed by the union of two molecules of gallic acid with the loss of one molecule of water. It is then digallic acid.

$$2C^7H^6O^5 \quad = \quad H^2O \quad + \quad C^{14}H^{10}O^9$$
Gallic acid.   Digallic acid.

The black mixture obtained by mixing solutions of tannin with ferric salts constitutes ink. A good black ink may be made by exhausting 100 grammes of powdered gall-nuts with 1.4 litres of water, and adding to the filtered liquid a solution of 50 grammes of gum arabic and 50 grammes of ferrous sulphate, each in the least quantity of water which will dissolve it. After stirring the mixture, it is allowed to stand exposed to the air until it becomes quite black.

The operation of tanning, or the conversion of animal skins into leather, depends on the formation in the skin of an insoluble compound of tannin and the albuminoid matter of the skin. The tannin is derived from oak bark, which is ground to a coarse powder and piled in alternate layers with the skins in deep vats. The vats are then filled with water, and the skins are allowed to soak for a few weeks or months, until they have become thoroughly penetrated by the tannin.

384. **Camphors.**—The highly aromatic solids that constitute the class of bodies called camphors are derived from a methyl-propyl-benzol called *cymene;* it is benzol in which two atoms of hydrogen have been replaced, one by a methyl group, CH^3, the other by a propyl group, C^3H^7. Its composition is therefore $C^{10}H^{14}$.

$$C^6H^6 \qquad\qquad C^6H^4(CH^3)(C^3H^7)$$
Benzol.   Cymene.

Cymene exists naturally in the essential oils of chamomile and thyme. It is a liquid having a pleasant odor. Its relations to the series of camphors are indicated in the following formulæ:

$C^{10}H^{14}$,   Cymene.
$C^{10}H^{14}O$,   Thymol, or thyme camphor.
$C^{10}H^{16}O$,   Camphol, or ordinary camphor.
$C^{10}H^{18}O$,   Borneol, or Borneo camphor.
$C^{10}H^{20}O$,   Menthol, or mint camphor.

385. THYMOL, $C^{10}H^{14}O$, is a phenol, one of the hydrogen atoms of the benzol group in cymene being replaced by the group OH. It exists in the essential oil of thyme, from which it may be extracted in the form of large, colorless, crystalline plates, fusible at 44°. It has a pleasant but penetrating odor, and is an excellent preservative, for it has antiseptic properties, destroying low forms of life.

386. CAMPHOL, $C^{10}H^{16}O$, is ordinary camphor, sometimes called laurel camphor, because it is obtained from the camphor laurel, a tree of China, Japan, and the Sunda Isles. It exists in all parts of the tree, but is extracted from the wood, which is chopped in small pieces and distilled with water. The camphor vapor condenses on rice-straw, with which the head of the still is filled: it is removed, and purified by a new sublimation. It forms semi-transparent, crystalline masses having a strong, aromatic odor and a sharp, burning taste. Its density at 0° is 1. It melts at 175°,

and boils at 204°; it is exceedingly volatile, and even at ordinary temperatures it sublimes in the vessels in which it is kept, condensing in the upper part in brilliant, colorless crystals. It is almost insoluble in water, but dissolves readily in alcohol and in ether. When small fragments of camphor are thrown on the surface of clean water, they move around with curious gyratory movements, which are caused by the pressure of the camphor vapor given off in unequal quantities from different parts of the surface of the fragments. The currents of •vapor may be made evident by dusting a small quantity of the fine powder called lycopodium on the surface of the water.

387. BORNEOL, $C^{10}H^{18}O$, is obtained from an aromatic tree of the Sunda Isles. It forms small colorless crystals, having an odor like that of camphor, but at the same time resembling that of pepper. It is insoluble in water, but dissolves in both alcohol and ether. Strong nitric acid converts it into ordinary camphor.

388. MENTHOL, $C^{10}H^{20}O$, is the solid part of the essential oil of mint, in which it is mixed with a hydrocarbon having the same composition as oil of turpentine. It forms colorless crystals, fusible at 36°.

389. **Indigo,** $C^{16}H^{10}N^2O^2$, is prepared from several indigo plants, which are cultivated principally in India. The leaves and stems of these plants are soaked in water for a day or two; a sort of fermentation takes place, after which the liquid is expressed and agitated in contact with the air. A blue deposit forms; it is collected and boiled with water in large copper vessels, and then drained, pressed, and broken up into the fragments in which indigo occurs in commerce. Indigo results from the decomposition of one of its compounds which exists in the plant.

The best indigo has a coppery appearance. It is not perfectly pure, but a small quantity may be purified by gently heating it in a small flask through which hydrogen is passed: the pure indigo, which is called *indigotin,* then sublimes and condenses in small crystals around the cooler portions of the flask. It is insoluble in water, alcohol, and ether, but dissolves in strong sulphuric acid, especially in fuming sulphuric acid. The dark-blue. solution so obtained is commonly called sulphate of indigo, and is used in dyeing.

390. WHITE INDIGO, $C^{16}H^{12}N^2O^2$.—When indigo is subjected to the action of reducing agents, such as sulphurous acid and hydrogen sulphide, it is converted into a dirty-white substance called white indigo. If a mixture of indigo, ferrous sulphate, and milk of lime be shaken in a corked bottle, and allowed to stand for a day or two, an alkaline solution of white indigo is obtained, from which the latter may be precipitated by a current of hydrochloric acid gas. The white indigo is insoluble in water, but dissolves in alcohol and in solutions of the alkaline hydrates. If a white cloth be dipped in the yellowish solution in the bottle, and then exposed to the air, it rapidly becomes blue. White indigo, which is a compound of hydrogen with indigo, is again converted into indigo on contact with the air, and the experiment with the cloth is an illustration of the manner in which in dyeing the insoluble blue indigo is deposited in the tissues of fabrics.

Indigo has been obtained artificially by a number of interesting reactions, which will ere long permit the manufacture of this important dye-stuff from the hydrocarbons of coal-tar.

LESSON XLVI.

NATURAL ALKALOIDS.

391. The compound ammonias, derived from ammonia by replacement of one or more of its hydrogen atoms by various groups or radicals, are powerful bases. They combine directly with acids, forming definite crystallizable salts. Thus, methyl ammonium chloride is as definite a body as ammonium chloride or potassium chloride.

KCl	NH^4.Cl	NH^3(CH^3).Cl
Potassium chloride.	Ammonium chloride.	Methylammonium chloride.

An immense number of compound ammonias have been formed,

and well studied, so that their molecular constitutions are perfectly known. Many plants contain principles which we have not been able to obtain artificially, but which so much resemble the compound ammonias in their chemical relations that we believe them to belong to the same class. They all contain nitrogen, and it is to the nitrogen atom or atoms that are due the basic properties of the compounds which are called natural *alkaloids;* that is, alkaline-like bodies. Most of these substances are poisonous; they all exert peculiar and active effects on the animal economy.

392. The processes adopted for the separation of the alkaloids from the plants or vegetable products in which they occur, vary according to the solubility of the particular alkaloid and its salts in various solvent agents. The alkaloids do not occur in an uncombined state in the plants, but united with some natural acid with which they form salts. If the natural salt be soluble in water, an aqueous extract of the compound may be used for the preparation of the alkaloid, but usually very dilute sulphuric acid is employed; sometimes the plant or product must be extracted with alcohol. The alkaloid is then set free by the addition of milk of lime or other alkaline hydrate, which will form a salt with the natural acid: sometimes the salt formed is insoluble in the liquid employed, while the alkaloid dissolves; sometimes it is the alkaloid which is insoluble and the salt which remains in solution. These circumstances must be investigated, and such a process adopted as will allow the alkaloid to be entirely separated, and it can then be easily purified by crystallization.

Two important natural alkaloids are liquid; they are conine and nicotine.

393. **Conine,** $C^8H^{15}N$, is the active principle of poisonous hemlock. It is extracted from the seeds, which are crushed and distilled with an alkaline hydrate. The conine distils, and is neutralized with sulphuric acid, which converts it into a sulphate, of which the solution is evaporated to a syrupy consistence, and then exhausted with a mixture of alcohol and ether. When the alcohol and ether have been evaporated, the conine sulphate is distilled with a strong solution of sodium hydrate, and sodium

sulphate is formed, while the conine set free condenses, together with a little water. The conine may be dried by calcium chloride. It is a colorless, oily liquid, having a disgusting odor. It is only slightly soluble in cold water, and still less soluble in hot water, but dissolves freely in alcohol and ether. It is very poisonous, as are also its salts.

394. **Nicotine,** $C^{10}H^{14}N^2$, exists in tobacco, probably in combination with malic acid. It may be obtained by extracting tobacco with boiling water, evaporating the filtered solution until it becomes a pasty mass, and mixing this residue with about twice its volume of alcohol. The alcoholic liquid separates in two layers, of which the upper contains the nicotine: it is decanted, and the alcohol distilled off. From the residue the nicotine is set free by potassium hydrate, and dissolved out by ether. The impure nicotine may then be converted into an oxalate by the addition of oxalic acid, and when this is decomposed by potassium hydrate, tolerably pure nicotine is obtained.

Nicotine is a colorless liquid, having an irritating and most penetrating odor. It is very soluble in water, alcohol, and ether. It boils between 240° and 250°. It is an energetic base, and is one of the most active poisons known. Tobacco contains from two to about seven per cent. of nicotine, the most esteemed varieties being those which contain the least.

395. **Theobromine,** $C^7H^8N^4O^2$, is the alkaloid of cacao, and may be extracted from cacao beans. It is a white, crystalline powder, having a bitter taste, and is not very poisonous.

396. **Caffeine,** $C^8H^{10}N^4O^2$, sometimes called theine, exists in coffee, tea, and several other plant products. It may be prepared by making a strong tincture of tea with cold alcohol, and precipitating the filtered liquid with basic lead acetate. The mixture is filtered, and freed from lead by a stream of hydrogen sulphide, after which it is again filtered, evaporated to a small volume, and while still hot is treated with potassium hydrate. Caffeine then crystallizes out as the liquid cools. It forms long, brilliant, white needles, containing one molecule of water of crystallization, which is driven out by a temperature of 100°. It has a bitter taste;

it is not very soluble in cold water, but dissolves readily in hot water and in alcohol.

397. **Morphine,** $C^{17}H^{19}NO^3$.—Opium, which is the thickened juice of the unripe capsules of the opium poppy, contains several alkaloids, of which the most important is morphine. The natural salts in which these alkaloids exist in opium are soluble in alcohol; laudanum and paregoric are tinctures of opium. Morphine may be most easily extracted by making a cold watery extract of finely-cut opium, evaporating the filtered liquid, and adding sodium carbonate to the still hot syrup. In the course of a day morphine deposits, and may be collected on a filter and dissolved in a little dilute acetic acid. The filtered solution is then decolorized by animal charcoal, and the morphine again precipitated by ammonia. Morphine is almost insoluble in water, and insoluble in ether. It is dissolved by hot alcohol, from which it separates in crystals containing one molecule of water. It has a very bitter taste.

When nitric acid is added to a little morphine, an orange-red color is produced. Ferric chloride solution produces a blue color with morphine. Morphine forms easily-crystallizable salts; the sulphate, chloride, and acetate are used in medicine; these salts are soluble in water.

The principal alkaloids of opium, besides morphine, are *codeine*, $C^{18}H^{21}NO^3$, and *narcotine*, $C^{22}H^{23}NO^7$; both are crystallizable solids. Codeine is morphine in which one hydrogen atom is replaced by methyl.

398. **Cocaine,** $C^{17}H^{21}NO^4$, exists in coca leaves, which are much used as a tonic and stimulant in South America.

399. **Atropine,** $C^{17}H^{23}NO^3$, is the alkaloid of belladonna or deadly nightshade, and is identical with *daturine*, the poisonous principle of stramonium, commonly called Jamestown weed. When it is administered internally, or applied to the eye, it produces dilatation of the pupil, which continues until all of the alkaloid has passed from the system. It is exceedingly poisonous.

400. **Quinine,** $C^{20}H^{24}N^2O^2$.—Cinchona bark, universally known and used as a remedy, contains several alkaloids, the more im-

portant being quinine and cinchonine.　These alkaloids are almost insoluble in water, and their sulphates are the forms in which they are principally employed in medicine.　For the manufacture of these salts, the bark is extracted with water acidulated with sulphuric acid, and the addition of milk of lime to the clear solution causes the precipitation of the insoluble alkaloids, mixed with calcium sulphate and the excess of lime.　The deposit is collected, dried, and exhausted with boiling alcohol, which dissolves the alkaloids.　When the filtered alcoholic solution is evaporated, the cinchonine crystallizes first, being least soluble, and the quinine is then neutralized with sulphuric acid, and the solution concentrated until the sulphate crystallizes.

Quinine sulphate crystallizes in very bitter, delicate white needles, only slightly soluble in cold water, but dissolving in about thirty times their weight of boiling water.　It dissolves readily in water containing a little free acid.　When ammonia is added to its solution, the free alkaloid quinine is precipitated as a white powder, while ammonium sulphate is formed.　Quinine is soluble in about its own weight of alcohol, and in twenty-two times its weight of ether.　It is almost insoluble in water.

401. Cinchonine, $C^{20}H^{24}N^2O$, is deposited from the alcoholic solution in which quinine still remains in solution during its extraction from cinchona bark.　Its properties much resemble those of quinine, from which, however, it may be distinguished by its insolubility in ether.　Quinine sulphate supposed to contain cinchonine sulphate is treated with a little ammonia-water, and then agitated with ether; any cinchonine present will remain undissolved.

402. Strychnine, $C^{21}H^{22}N^2O^2$.—The poisonous and medicinal properties of nux vomica are due principally to two alkaloids, strychnine and brucine.　They are almost insoluble in water, and may be extracted from nux vomica by a process like that which serves for the separation of quinine.　They are both exceedingly bitter, crystallizable solids, nearly insoluble in water.　Strychnine is almost insoluble in alcohol and ether, but dissolves in chloroform.　Brucine is soluble in alcohol, and somewhat soluble in ether.

If a small fragment of potassium dichromate is placed beside a crystal of strychnine, and both are touched with a drop of sulphuric acid, a rich blue color is produced, which quickly changes to violet, purple, and red, and finally fades.

Strychnine is a violent poison; when taken even in comparatively trifling quantities, it produces terrible convulsions, resembling those of tetanus.

LESSON XLVII.

METALS.—SPECTRUM ANALYSIS.

403. The classification of the elements as metals and nonmetals is more for the sake of convenience than for the indication of absolute properties of either class. We may, however, consider that certain general properties are peculiarly manifested by the metals: they are good conductors of heat and electricity; they are capable of acquiring a brilliant lustre, which is called the metallic lustre. These properties are, however, more or less developed in some of the elements which we have already studied. It is not so, however, with a chemical property: the metals are capable of replacing the hydrogen of the oxygen acids, forming salts. Some of these salts we have already studied, and we have seen how the combining power or worth of a metallic atom is indicated by the number of hydrogen atoms which it is able to replace in an acid. Yet even in this respect the metals and nonmetals do not seem to be widely separated, for antimony, which is so closely related to phosphorus and arsenic by the compositions and chemical natures of its compounds, is also capable of forming a few salts.

The physical properties of the metals are most varied. They are opaque; but many of them can be reduced to sheets so thin that they allow the passage of a faint light whose color depends on the metal employed. Their densities vary from 0.59, that of lithium, to 22.4, of osmium; their freezing points, from 39° below

L q 21

0°, where mercury freezes, to about 2500°. Some, like manganese and chromium, are hard enough to scratch glass; others are soft enough to be scratched and even cut by the finger-nail, like potassium, sodium, and lead. Most of the metals are malleable and ductile; they can be beaten or rolled into sheets and drawn into wires. All the metals are insoluble in water.

404. **Natural State of the Metals.**—The condition in which the metals are encountered in nature depends upon the other elements for which they have strong affinities. Some of them are often found in the metallic state: among these are gold, silver, copper, and bismuth; they are then called *native metals.*

In general, the elements which are more usually combined with the metals in their ores are oxygen, sulphur, and chlorine. Iron, zinc, and manganese are found as oxides; iron, copper, lead, mercury, zinc, and silver, as sulphides; sodium and silver, as chlorides; calcium and magnesium, as carbonates; aluminium, as silicate.

The process adopted for the extraction of a metal must of course depend upon the nature of its ores: oxides are heated with charcoal; carbonates are first heated to drive off carbon dioxide, and the resulting oxide is reduced by charcoal. Sulphides are roasted,— that is, heated in the air,—by which the sulphur is converted into sulphur dioxide, and passes off in that gas, while an oxide of the metal is formed. The methods employed for the reduction of the chlorides differ according to the metal.

405. **Alloys** are the compounds or mixtures which the metals form with one another. In the molten state, many of the metals are capable of mixing with one another in all proportions; but by certain precautions, and the use of the proper proportions of metals, many alloys become crystallizable, and assume the properties of true chemical compounds: such compounds, of course, contain their respective metals in the proportions required by the atomic weights. The alloys of mercury are called *amalgams.*

SPECTRUM ANALYSIS.

406. When a ray of white light is passed through a prism, it is dispersed or separated into a spectrum consisting of all the

colors, from red to violet. If the ray be narrow and rectangular, such as is obtained by excluding all light except that which passes through a rectangular slit, and the spectrum be thrown on a white screen, the colors will not be confused, nor will they be distinctly separate, but will blend gradually from red to orange, yellow, green, blue, indigo, and violet. When any solid substance is heated to bright incandescence, it emits a white light, whose spectrum will contain all the prismatic colors. We have already had occasion to observe the dazzling whiteness of the combustion of magnesium and phosphorus. We have seen, also, that an alcoholic solution of boric acid burns with a green flame. We may prepare alcoholic solutions of sodium chloride, strontium chloride, and barium chloride, by shaking those substances in separate bottles with alcohol which is not too strong. When we burn these solutions, we find that the flame is colored yellow by the sodium salt, red by that of strontium, and green by that of barium, small quantities of the salts being carried into the vapor of alcohol, and volatilized by the high temperature of the flame. If on the end of a small platinum wire we introduce separately a little of each of these salts into the flame of an alcohol lamp, or, better, that of a Bunsen burner, we find that the same coloration is produced. If the light from such a flame be passed through a narrow slit, and then through a prism, we find that the color is invariable for each substance. The sodium salt produces not only a yellow light, but a particular shade of yellow, which, because it has passed through the straight slit, forms a peculiar spectrum, consisting of a single line of yellow light. If we keep the slit and prism in the same positions, it matters not what compound of sodium we introduce into the flame, the same yellow line is always produced, and on the same part of the screen. If we introduce a little lithium chloride into the flame, the latter will be colored red, and we will find on the screen a red line of a fixed and constant shade, and always in the same position.

Analogous facts have been discovered for all the elements, and we may say, generally, that while the light emitted by an incandescent solid depends upon the temperature, being first dull red, then orange, yellow, and white, an incandescent gas or vapor, on the contrary, always emits light of a constant color, depending on the nature of the substance. Usually the spectrum of an element does not consist of a simple line, one color only, but of several and sometimes many lines ; but in each case the spectrum is peculiar to the element.

407. These principles have been applied in spectroscopic analysis for the detection of the elements, and the instrument employed is called a spectroscope. It consists of a narrow slit at the end of a metallic tube containing a lens (A, Fig. 100), by which

FIG. 100.

the rays of light are made to enter a prism in parallel lines: the light, having passed through the prism, is directed into a short telescope (B), by which the rays are again brought to a focus, so that the image may be examined by the eye. In order that the exact position of any line may be accurately observed and the line identified, the image of a small graduated scale illuminated by a faint light (C) is reflected into the telescope from the side of the prism opposite to the slit. This scale corresponds to a similar graduated scale which we might make. on a screen on which a spectrum is thrown. The substance of which the spectrum is to be examined is then heated on a platinum wire in a Bunsen-burner flame exactly opposite the slit. Sometimes electric sparks are passed between points of the substance, and if the latter be a gas it must be enclosed in a tube and rendered luminous by sparks passed through it from an induction coil. The spectra of most metals are very brilliant lines (see frontispiece), so brilliant that they entirely obscure the more faint and broader bands of the spectra of the non-metals; for this reason, when we heat sodium chloride in the burner flame, we can only observe the spectrum of sodium by the spectroscope, although the chlorine must also produce its spectrum.

Spectroscopic analysis is exceedingly delicate: $\frac{1}{3,000,000}$ of a milligramme of sodium chloride introduced into the burner flame will cause the yellow sodium line to flash out for an instant. While studying spectra, several chemists have observed lines which were not produced by any substance then known, and have thus been led to the discovery of new elements, of which small quantities were present in the substances under examination.

The study of the spectrum of the sun's light and the light of the stars has shown us perfectly the elements which exist in an incandescent state in the atmospheres of those far-distant bodies. Some lines in the spectrum of sunlight corresponded to those of no element known on the earth, and chemists concluded that this unknown substance existed in the sun's atmosphere, and named it helium. This same element has recently been discovered, by the aid of spectrum analysis, in the lava from Vesuvius,—another evidence of the distribution of the same elements throughout the universe, and of the unique source of all matter.

LESSON XLVIII.

METALLIC COMPOUNDS.—SPECIFIC HEAT.

408. Before we undertake the study of the individual metals, we will pass in review some of the facts which we have already learned concerning metallic compounds, and will develop them by the consideration of new details.

Oxides and Hydrates.—All excepting a few of the metals combine directly with oxygen at various temperatures. Potassium is the only metal which is oxidized by cold dry air, and for the oxidation of some metals a very high temperature is required. The number of atoms of oxygen and of metal which combine together depends on the atomicity of the metal. Two atoms of a monatomic metal combine with one atom of oxygen, while in the formation of a monoxide only one atom of a diatomic metal takes part. The oxides of lithium, sodium, and potassium are soluble in water, but in dissolving they form hydrates, which we must admit contain a hydroxyl group.

$$K^2O \quad + \quad H^2O \quad = \quad 2KOH$$

The hydrates of these three metals are the alkaline hydrates, the metals being called the alkaline metals. Nearly all the oxides are capable, under certain conditions, of forming hydrates, containing one or more hydroxyl groups, and the oxidation or rusting of metals in moist air always results in the formation of hydrates and not oxides. Calcium, strontium, and barium oxides (or hydrates) are less soluble in water than those of lithium, sodium, and potassium. The other oxides are almost or entirely insoluble in water.

Some metals form several compounds with oxygen, and those which contain the least oxygen are basic oxides, or bases; they are capable of reacting with acids, forming water and salts in which the hydrogen of the acid is replaced by metal. Most of the oxides containing two atoms of oxygen also react with acids, but, while water and a salt are formed, an atom of oxygen is disengaged. In this manner sulphuric acid liberates ozone from barium dioxide.

$$3BaO^2 \quad + \quad 3H^2SO^4 \quad = \quad 3BaSO^4 \quad + \quad O^3 \quad + \quad 3H^2O$$
Barium dioxide.   Barium sulphate.   Ozone.

The sesquioxides, those containing two atoms of metal and three of oxygen, form salts which are important to understand. Ferric oxide has the composition Fe^2O^3: when it reacts with acids, the two atoms of iron always enter into one molecule of the salt formed. The vapor-density of ferric chloride shows that its molecule contains Fe^2Cl^6: it is formed by the action of six molecules of hydrochloric acid on one molecule of ferric oxide.

$$Fe^2O^3 \quad + \quad 6HCl \quad = \quad Fe^2Cl^6 \quad + \quad 3H^2O$$

We believe that in ferric oxide and ferric chloride the atoms of iron are tetratomic, and that two atoms combine together, forming a hexatomic couple, just as two atoms of carbon are related in the hydrocarbon C^2H^6. We may then contrast the molecular structure of ferric oxide with that of arsenious oxide or phosphorous oxide: each molecule contains three atoms of oxygen and two atoms of another element, but these atoms are very differently arranged.

O=Fe–Fe=O   O=As–O–As=O   O=P–O–P=O
    \ /
    O
Ferric oxide.   Arsenious oxide.   Phosphorus trioxide.

Ferric oxide, then, forms ferric salts in which the two atoms of metal replace six atoms of hydrogen of the acids. There are many other sesquioxides of the same nature. At the same time these oxides have in some cases the curious property of acting like acid radicals. Ferrous oxide, FeO, is an energetic basic oxide; it combines with ferric oxide, and the combination, $Fe^3O^4 = FeO.Fe^2O^3$, called ferroso-ferric oxide, constitutes magnetic iron ore, or black oxide of iron: it is called a saline, or salt-like oxide.

The oxides containing one atom of metal combined with three or more atoms of oxygen correspond to metallic acids. They are capable of reacting with water or with basic oxides, forming well-marked acids and salts. Chromium trioxide, CrO^3, corresponds to chromic acid, $H^2CrO^4 = H^2O + CrO^3$.

When highly heated with charcoal or in a current of hydrogen, most of the metallic oxides are reduced to metal, while either carbon monoxide, carbon dioxide, or water is formed. The oxides of calcium, barium, strontium, magnesium, aluminium, potassium, sodium, and lithium are not reduced by hydrogen, and the first five are not reduced by carbon.

409. Sulphides.—Nearly all the metals combine directly with sulphur at certain temperatures, and the sulphides formed are analogous in composition to the oxides. The alkaline sulphides, and those of calcium, strontium, and barium, are soluble in water; the others are insoluble.

At temperatures depending upon the nature of the metal and the state of division of the sulphide, oxygen decomposes all the sulphides, sometimes forming sulphur dioxide and leaving a metallic oxide or even the free metal, sometimes oxidizing the sulphide to sulphate, according to the nature of the metal. If a mixture of potassium sulphate and powdered charcoal be heated to redness in a covered crucible, a porous black mass is obtained; it contains potassium sulphide, and if it be broken up and thrown into the air, this sulphide is oxidized to potassium sulphate, producing a shower of sparks.

$$K^2S \quad + \quad 2O^2 \quad = \quad K^2SO^4$$
$$\text{Potassium sulphide.} \qquad\qquad \text{Potassium sulphate.}$$

410. Chlorides, Bromides, and Iodides.—Excepting platinum, all the metals combine directly with free chlorine; since in its compounds with the metals, as in its compound with hydrogen, chlorine is a monatomic element, the number of chlorine atoms contained in a molecule of a metallic chloride is an indication of the atomicity of the metal.

All the chlorides are soluble in water, excepting silver chloride, mercurous chloride, and cuprous chloride: plumbic chloride is only slightly soluble.

As a rule, the bromides are more soluble than the corresponding chlorides, and the iodides more soluble than the bromides.

411. The color of the metallic compounds may be remembered by certain general principles. If both the corresponding oxide or hydrate and the corresponding acid be colorless, the salts are also colorless. If either the acid or the oxide or hydrate be colored, the salts are colored. The salts formed by the same metal with colorless acids are of about the same color: with colorless oxides or hydrates the same colored acid forms corresponding salts of about the same color. In many cases the color of metallic compounds depends on water of crystallization, as we have already seen (§ 55), and is lost when that water is expelled.

SPECIFIC HEAT.

412. The atomic weights of the metals cannot often be estimated from their vapor-densities, for many of them are volatile only at such high temperatures that it is impracticable, or even impossible, to determine the densities of their vapors. Some of the metals form volatile compounds with chlorine or with various hydrocarbon radicals; and since the molecular weights of these compounds can be determined without difficulty from the densities of their vapors, we can arrive at the atomic weight of the corresponding metal.

The compounds of the metals with oxygen, with chlorine, and with other bodies of course contain a fixed number of atoms of metal with a definite number of atoms of other elements of which the atomic weight is known. Thus, we know that for every sixteen parts of oxygen, potassium oxide contains 78.2 parts of potassium. We have already studied the reasoning by which we conclude that the atomic weight of oxygen is sixteen; how shall we determine whether the 78.2 parts of potassium represent one, two, or three atoms of that metal?

In order to raise the temperatures of equal weights of different substances through the same number of thermometric degrees, very different quantities of heat are required. If we expose one kilogramme of mercury and one kilogramme of water, both at 0°, to the same source of heat, we find that when the water will have been heated to 1° the mercury will be at 30°. If, on the other hand, we place one kilogramme of mercury at 100°, with some ice, in a vessel so constructed that all of the heat will be employed in melting the ice, we find that only one-thirtieth as much ice will be melted as if we put in the same vessel one kilogramme of water at 100°. The relative quantities of heat which are required to raise equal weights of different substances through the same number of thermometric degrees, are called the *specific heats* of the substances. Water is the substance whose specific heat is chosen as unity, and the specific heat of any substance then represents the quantity of heat required to raise a given weight of the substance through one degree, compared with that which will raise the same weight of water through the same temperature. The specific heat of mercury is, then, $\frac{1}{30} = 0.03333$. On comparing the specific heats of the liquid or solid elements, it has been found that just in the same proportion that the atomic weight increases, the specific heat diminishes; the specific heats are inversely as the atomic weights. The product of the specific heat of any liquid or solid element by its atomic weight should, then, always give the same figures. This important fact was discovered by Dulong and Petit, and is generally called Dulong and Petit's law: its import is evidently that the atoms of the different elements all possess the same specific heat. An examination of the figures expressing the quantities involved will show the facts on which the law is based:

NAME OF ELEMENT.	ATOMIC WEIGHT.	SPECIFIC HEAT.	PRODUCT.
Lithium.	7	0.9408	6.586
Boron	11	0.5	5.5
Carbon	12	0.46	5.52
Sodium	23	0.2934	6.748
Magnesium	24	0.2499	5.998
Aluminium	27	0.2143	5.786
Phosphorus	31	0.1887	5.850
Sulphur	32	0.2026	6.483
Potassium	39.1	0.1695	6.500
Zinc	65.2	0.0955	6.230
Bromine	80	0.0843	6.744
Iodine	127	0.0541	6.873
Mercury	200	0.0325	6.494

The average of the products of the atomic weights by the specific heats is 6.4 : however, while the product, which we may call the atomic heat, is always near the number 6.4, it varies within certain limits. Were it always 6.4, we could readily obtain the atomic weight of any element by dividing 6.4 by the specific heat; as it is, the figures expressing the specific heat enable us to choose between two numbers widely separated, and have in several cases indicated that the number which had been supposed to represent the atomic weight should be halved or doubled.

LESSON XLIX.

LITHIUM.—SODIUM.—POTASSIUM.

413. **Lithium,** Li = 7.—The metal lithium is very widely diffused in nature, but is found only in small quantity, excepting in Bohemia, where there are large mountains of a peculiar lithium mica, called *lepidolite,* containing from three to six per cent. of lithium. Mica is a silicate of aluminium, potassium, and magnesium or iron; in lepidolite a proportion of the potassium is replaced by lithium.

Metallic lithium is obtained by decomposing fused lithium chloride, LiCl, a colorless soluble salt, by a current of electricity. It is a silver-white metal, and does not tarnish in dry air. Its density is the lowest of any solid known, being about 0.58. It melts at 180°, and may be melted in contact with the air without becoming oxidized : when heated to redness in the air or in oxygen, it burns with a dazzling white flame. Lithium soon becomes tarnished in moist air, being converted into lithium hydrate, LiOH ; when it is thrown on the surface of water, the same hydrate is formed, the water being decomposed and hydrogen disengaged. The lithium salts are soluble in water, and are colorless unless the corresponding acid is colored. They communicate a red color to the Bunsen-burner flame, and their spectrum is characterized by a brilliant red and a more faint orange line (see frontispiece).

414. **Sodium,** Na = 23.—Nearly forty per cent. of the immense quantities of sodium chloride existing in the ocean, in deposits of rock-salt, and in salt wells, consists of sodium. We have already studied the processes by which sodium chloride is converted into sodium carbonate. It is from the latter compound that the metal is manufactured. The sodium carbonate is thoroughly dried, and

FIG. 101.

mixed with charcoal and enough lime to prevent the mixture from melting, for if it melted the charcoal would float on the liquid carbonate. The mixture is then heated to a very high temperature in large cast-iron cylinders. Carbon monoxide is disengaged, and the sodium vapor condenses in flat receivers, from which the liquid metal runs into vessels containing naphtha, or light coal-oil (Fig. 101).

Sodium is a white metal, so soft that it can easily be cut like wax. It is a little lighter than water, its density being 0.97. It melts at 90.6°, and boils at a red heat. It can be melted in the air without taking fire. Its bright surface rapidly tarnishes in moist air, being converted into sodium hydrate. It is preserved in bottles containing naphtha, by which it is protected from the air. When a small piece of sodium is thrown on water, chemical action at once begins; the sodium melts and rushes about with a hissing noise. The reaction frequently terminates with an explosion by which small particles of sodium hydrate are thrown out, and we must make the experiment at a safe distance from the eyes. If the motion of the sodium be arrested, the heat will accumulate

FIG. 102.

sufficiently to ignite the escaping hydrogen. We float a piece of filter-paper on some water in a plate, and throw on this wet paper a small piece of sodium: it at once melts, and soon the hydrogen takes fire, burning with a flame tinged bright yellow by a little sodium vapor (Fig. 102).

415. SODIUM HYDRATE, NaOH, is the product of the reaction of sodium with water. It is manufactured by a number of processes: when sodium carbonate in rather dilute solution is boiled with milk of lime, sodium hydrate passes into solution, while insoluble calcium carbonate is formed.

$$Na^2CO^3 + Ca(OH)^2 = 2NaOH + CaCO^3$$
Calcium hydrate.

This operation is somewhat expensive, on account of the large quantity of water which must be boiled away from the sodium hydrate. The Le Blanc process for the manufacture of sodium carbonate (§ 240) can with slight modifications be made to yield considerable quantities of an impure sodium hydrate, which remains in solution after the sodium carbonate has crystallized.

Much sodium hydrate of an excellent quality is now manufactured from cryolite (§ 240). The powdered mineral is boiled with milk of lime, insoluble calcium fluoride and a solution of aluminate of sodium being obtained.

$$Al^2Fl^6.6NaFl + 6Ca(OH)^2 = 6CaFl^2 + Al^2O^3.3Na^2O + 6H^2O$$
Cryolite.   Calcium hydrate. Calcium fluoride. Aluminate of sodium.

The filtered solution is then boiled with a new quantity of pulverized cryolite, and all the sodium is so converted into soluble sodium fluoride, while aluminium oxide is precipitated.

$$Al^2O^3.3Na^2O \quad + \quad Al^2Fl^6.6NaFl \quad = \quad 2Al^2O^3 \quad + \quad 12NaFl$$

When the insoluble aluminium oxide has settled, the clear solution of sodium fluoride is drawn off and boiled with milk of lime: calcium fluoride is precipitated, while sodium hydrate remains in solution.

$$2NaFl \quad + \quad Ca(OH)^2 \quad = \quad 2NaOH \quad + \quad CaFl^2$$

The aluminium oxide obtained in this operation is used for the manufacture of alum, and the calcium fluoride is employed as a flux, or fusing agent, in the separation of many metals from their ores.

By whatever process it be obtained, the solution of sodium hydrate is evaporated to dryness, and subsequently fused in iron boilers out of contact with the air. It then forms a hard, white solid, which if left exposed to the air absorbs moisture and carbon dioxide, becoming converted into sodium carbonate. It is very soluble in water, and very caustic. It is commonly known as concentrated lye, and is employed in enormous quantities for the manufacture of soap.

When fragments of sodium are thrown into sodium hydrate melted in an iron dish, hydrogen is disengaged, and sodium oxide is formed.

$$2NaOH \quad + \quad Na^2 \quad = \quad 2Na^2O \quad + \quad H^2$$

416. SODIUM CHLORIDE, $NaCl$.—This compound is common salt. It exists in numerous and immense deposits of rock-salt in many localities. It is found in salt wells and salt springs, and constitutes the greater portion of the solid matter of sea-water. The water of the Atlantic Ocean contains, according to the locality, from 32 to 38 grammes of solid matter per litre; the water of the Pacific contains somewhat less, but the average proportion of common salt in each is about thirty grammes per litre. The other constituents of sea-water are principally chlorides and sulphates of potassium, magnesium, and calcium, with small quantities of bromides and iodides. When the water is evaporated, the sodium chloride separates first, while the other salts remain in more concentrated solution. In warm countries the evaporation is often accomplished by the heat of the sun and exposure to con-

stant winds, in large shallow basins into which the water is either pumped or led by sluices from the sea.

Sodium chloride crystallizes in cubes, which may be obtained of large dimensions and perfectly transparent, by the slow evaporation of a saturated solution. ˜It is anhydrous, but the crystals, especially if small, usually retain in the spaces between them a small quantity of water, which is converted into steam and causes the crystals to decrepitate—that is, crack into small pieces—when they are heated. It is soluble in less than three times its weight of cold water, and in about two and a half times its weight of boiling water. It is insoluble in pure alcohol. It melts when heated to redness, and volatilizes at a higher temperature.

417. Tests for Sodium.—Since all the ordinary sodium salts are soluble and colorless, none of the ordinary reagents produce either precipitates or colors in their solutions. Hydrofluosilicic acid yields a white precipitate of silico-sodium fluoride (§ 221). We may readily recognize the presence of sodium by the yellow color which all its compounds communicate to the colorless flame of the Bunsen burner.

418. Potassium, K = 39.—For a long time the principal source of potassium was the potassium carbonate obtained from wood-ashes, and from that substance the other potassium compounds were manufactured. Large quantities of potassium carbonate are now obtained from the double chloride of potassium and magnesium, called, from its source, Stassfurth salt (§ 242).

Metallic potassium is prepared by a process exactly like that which yields sodium; potassium carbonate is heated with charcoal; a higher temperature is required than for the preparation of sodium. Potassium occurs in commerce as round, brownish masses, kept under naphtha for the same reason that sodium is so preserved. It is quite soft, and yields readily to the pressure of the finger-nail. When freshly cut, it displays a brilliant surface, but this rapidly tarnishes by the action of the air. Its density is about 0.86 : it melts at 62.5°, and boils at a red heat, emitting a green vapor. It combines with dry oxygen, and in the cold, forming potassium oxide, K^2O. In moist air it is converted into the hydrate KOH. When a small piece of potassium is thrown in water, it decomposes the latter so violently that the hydrogen disengaged is at once ignited, and the potassium rushes about in the burning gas, whose flame is tinged violet by the metal (Fig. 103). The experiment termi-nates with a little explosion, for the globule of potassium hydrate formed is at a very high temperature, and when it cools sufficiently to come in contact with the water, there is a sudden for-mation of steam.

FIG. 103.

419. POTASSIUM HYDRATE, KOH, is prepared by boiling milk of lime with a rather dilute solution of potassium carbonate. As soon as the reaction has terminated, the solution of potassium hy-drate is poured off the deposit of insoluble calcium carbonate, and is rapidly evaporated to dryness in iron or silver dishes. It is then fused, and cast in cylindrical moulds (Fig. 104), so that it

FIG. 104.

usually occurs in commerce in round sticks. It commonly contains considerable quantities of lime, potassium carbonate, silicate, and other salts. It may be purified by dissolving it in alcohol in which only the hydrate is soluble, decanting the clear solution, and fusing in a silver dish the residue from which the alcohol has been distilled. It is white and opaque, and has a density of 2.1. It melts at a red heat, and volatilizes at a higher temperature. It is exceedingly soluble in water, and, when exposed to the air, absorbs moisture and carbon dioxide, deliquescing to a liquid consisting of a solution of the carbonate. It is very caustic and corrosive, rapidly destroying animal tissues. It is employed in making soft soap.

420. POTASSIUM CHLORIDE, KCl.—This salt forms transparent, colorless cubes, exactly resembling the crystals of sodium chloride. It is found native in some localities, and, in combination with magnesium chloride, constitutes Stassfurth salt, $KCl,MgCl^2$ $+ 6H^2O$ (see § 242). It dissolves in about three times its weight of cold water, and in less than twice its weight of boiling water.

421. POTASSIUM BROMIDE, KBr, is employed extensively in medicine. It is usually made by adding to bromine enough strong solution of potassium hydrate to almost decolorize the liquid. The reaction yields a mixture of potassium bromide and potassium bromate.

$$6KOH \quad + \quad 3Br^2 \quad = \quad 5KBr \quad + \quad KBrO^3 \quad + \quad 3H^2O$$

The mixture is evaporated to dryness, and then heated to redness, sometimes with the addition of a little powdered charcoal; the bromate then loses its oxygen, and is converted into bromide. After cooling, the mass is dissolved in water, and the salt made to crystallize. Potassium bromide forms beautiful colorless cubes, having an intensely salty taste, and soluble in about one and a half times their weight of cold water.

422. POTASSIUM IODIDE, KI, is prepared in exactly the same manner as the bromide, iodine being substituted for the bromine. It also crystallizes in colorless cubes having a salty and at the same time bitter taste. It dissolves in about two-thirds its weight of cold water, and the solution will dissolve large quantities of iodine, becoming dark brown in color. Both the bromide and iodide of potassium of commerce occur not in transparent but in white, opaque crystals: they contain a trace of free alkali. When the transparent crystals have been put in the market, they have found no sale, being supposed to be impure.

423. TESTS FOR POTASSIUM.—Like the salts of sodium, most of the potassium salts are colorless and soluble, and their solutions

are neither precipitated nor colored by the ordinary reagents. Hydrofluosilicic acid produces a gelatinous white precipitate of silico-potassium fluoride. When the solution of a potassium salt is mixed with a strong solution of tartaric acid, a white crystalline precipitate of cream of tartar soon separates. Platinic chloride, $PtCl^4$, produces a yellow, crystalline precipitate of potassio-platinic chloride, $(KCl)^2PtCl^4$. The potassium compounds impart a violet color to flame, but the color is rather delicate, and often masked by the presence of sodium or lithium: it is then examined through a blue glass which does not allow the passage of the light from the sodium and lithium flames, but through which the violet potassium flame is distinctly visible.

424. **Analogies of Lithium, Sodium, and Potassium.**—When we compare together the compounds of the metals which we have just studied, we find that the three form a group presenting the most evident chemical analogies. They are monatomic metals, capable of replacing the hydrogen of acids, atom for atom. One atom of either metal will combine with one atom of chlorine, or with one hydroxyl group, but two atoms are required to combine with the diatomic atom of oxygen. Moreover, the corresponding salts of these metals are isomorphous: they crystallize either in exactly the same forms, or in forms which are easily derived one from the other. The rare metals cæsium and rubidium form part of the group just considered.

LESSON L.

SILVER. Ag = 108.

425. Silver is found in the metallic state, and in combination with many other elements, among the more ordinary of which are sulphur, chlorine, arsenic, antimony, and lead.

When the silver ores do not contain lead, the silver is extracted by amalgamating it with mercury and then driving off the latter by the action of heat. Several processes are employed; in all of them the silver is first converted into silver chloride. The German method consists in roasting the powdered ore with common salt: the sulphides present are thus oxidized, while the silver is converted into chloride. The cold mass is pulverized, and washed with water to remove all soluble salts formed; the residue is then put into barrels with water and scrap iron, and these amalgamation

barrels are rotated by machinery until the contents are thoroughly mixed (Fig. 105). Silver is set free, while the chlorine combines with the iron. Mercury is now introduced, and forms an amalgam with the silver. The liquid amalgam is strongly pressed in canvas bags, and the greater part of the mercury is squeezed out. The semi-solid amalgam remaining is

FIG. 105.

heated until the mercury is expelled, and the residue is metallic silver containing a certain proportion of copper derived from copper sulphide in the ore.

In the process adopted on the Pacific slope, the ore is reduced to a very fine powder, which is mixed with a proportion of common salt depending on the amount of silver to be chloridized. By appropriate machinery, this mixture is thrown into a tall chimney-shaft through which a current of very hot air is rising. Under these circumstances, all the silver is at once converted into chloride, which falls to the bottom of the shaft, from which it is removed when about a ton has accumulated. It is then washed in a stream of water, and the insoluble silver chloride settles as a pulpy mass. This pulp is mixed with a little cupric sulphate and common salt in iron pans heated by steam, and about one hundred and fifty pounds of mercury are added for every ton of the pulp. After five or six hours' grinding, the mercury contains all the silver, which is reduced partly by the iron of the pan, partly by the conversion of some mercury into chloride. The amalgam is then agitated with water, and, after it is dried, the mercury is driven off by distillation in cast-iron retorts.

426. Galena, or lead sulphide, an important lead ore, often contains a considerable proportion of silver, which forms an alloy with the lead when the ore is reduced. Large quantities of silver are

extracted from such lead by a process called, from the name of its inventor, Pattinsonizing. When a melted alloy of lead and silver containing even small quantities of the latter metal is allowed to cool, almost pure lead first solidifies in crystals; this is the fact on which the process is based. The molten lead is allowed to cool slowly, and, by means of large ladles, the crystals of lead are removed as fast as they are formed, so that the metal which remains liquid to the last is an alloy rich in silver (Fig. 106).

FIG. 106.

As the lead crystals so removed still contain a little silver, they are submitted a second and a third time to the same operation, so that pure lead is obtained on one hand, and a very rich silver alloy on the other. The lead is entirely removed from the alloy by a process called cupellation. The metal is melted on a shallow hearth swept by the flame of a small furnace. This hearth, which is called a cupel, is covered by a sheet-iron dome (G, Fig. 107), which can be raised and lowered as necessary. When the whole

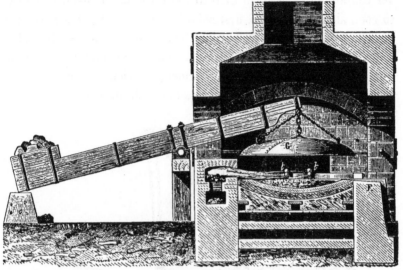

FIG. 107.

of the metal is melted, a blast of air is blown on its surface from pipes called tuyères (t t), and the lead is oxidized. The oxide melts, and, being lighter than the metal, is drawn off through a notch cut in the side of the cupel, and the notch is gradually deepened as the level of the fused metal becomes lowered. The silver does not oxidize, and at last, when its surface is covered with only a thin layer of molten lead oxide, that layer breaks suddenly, and the brilliant surface of the silver appears with a flash. The blast of air is then stopped, and the silver is either drawn off into ingot-moulds or allowed to solidify in the cupel.

427. Silver is the most brilliantly white metal. It is exceedingly malleable and ductile. Its density is 10.5. It does not tarnish on exposure to the air, but above its melting point, which is about 1000°, it absorbs or combines with about twenty-two times its volume of oxygen from the air. The oxygen is expelled violently as the metal solidifies, and portions of the still liquid silver are often projected from the vessel, while its surface is thrown into curious tree-like forms. This phenomenon is called spitting.

Ozone oxidizes silver to the dioxide Ag^2O^2. It is blackened by hydrogen sulphide, silver sulphide being formed on its surface; the discoloration of silver-ware is due to traces of hydrogen sulphide in the air; the sulphur in eggs, mustard, etc., rapidly blackens silver spoons. Boiling sulphuric acid dissolves silver slowly, converting it into sulphate; hydrochloric acid forms insoluble silver chloride on its surface, and the metal beneath is so protected from further action. It dissolves readily in nitric acid, red vapors being disengaged and silver nitrate formed. It is not attacked by the alkaline hydrates, and therefore silver vessels are used for the concentration and fusion of those compounds.

428. SILVER CHLORIDE, AgCl, is one of the more important silver ores; it is the mineral *horn-silver*, so called from its appearance and somewhat elastic, horn-like structure. We have already seen that it is precipitated on the addition of hydrochloric acid or a soluble chloride to solution of silver nitrate. It then forms a white, curdy precipitate, which darkens and undergoes partial decomposition on exposure to light. If a piece of zinc be placed in some recently-precipitated and still moist silver chloride, the whole of the silver soon separates in the form of a gray powder, while zinc chloride is formed. Pure silver may be thus obtained, but for that purpose the silver chloride should be previously well washed with dilute sulphuric acid, and the silver powder must be thoroughly washed by shaking it many times with water and then allowing it to settle. Pure silver may also be made by fusing the well-washed chloride with sodium carbonate; carbon dioxide and oxygen are disengaged, sodium chloride is formed, and the silver remains as a button at the bottom of the crucible. When recently precipitated, silver chloride dissolves readily in ammonia-water, from which it is again deposited when the ammonia is neutralized by an acid.

429. SILVER OXIDE, Ag^2O, is made either by precipitating a solution of silver nitrate by potassium hydrate, or by boiling well-washed silver chloride with potassium or sodium hydrate solution. It is a brown powder, insoluble in water, and decomposed by heat into silver and oxygen.

430. SILVER SULPHIDE, Ag^2S, is found native in small octahedral crystals. It is precipitated by the action of hydrogen sulphide on solution of silver nitrate, and may be formed by the direct union of silver and sulphur at a slightly-elevated temperature.

431. TESTS FOR SILVER.—In solutions of silver salts, hydrochloric acid produces a white precipitate of silver chloride; this precipitate is soluble in ammonia-water, and darkens in color when exposed to light. Potassium iodide solution gives a yellow precipitate of silver iodide, AgI, which also darkens by the action of light, but is only slightly soluble in ammonia. Hydrogen sulphide precipitates black silver sulphide. Potassium chromate precipitates red silver chromate, Ag^2CrO^4, in neutral solutions which are not too dilute.

432. SILVER-PLATING.—It is often desired to cover other metals or glass with a thin layer of silver. This may be accomplished in several manners. Copper objects may be silvered by rubbing them with a mixture of moist silver chloride and sodium carbonate, but the layer of silver so deposited is very thin. The metals are most readily and evenly silvered by connecting the object to be plated with the zinc pole of a voltaic battery and immersing it in a solution of silver and potassium double cyanide, made by boiling silver chloride in a solution of potassium cyanide. The positive pole of the battery is connected with a plate of silver, or silver coin, immersed in the same liquid. The silver solution then always retains its strength, for the metal dissolving from the positive electrode replaces that which is deposited on the article to be silvered. We may readily coat the interior of a test-tube with a thin layer of silver by pouring into it a solution of silver nitrate and sufficient ammonia-water to redissolve the precipitate first formed : we then add a few drops of a solution of tartaric acid, and place the tube in water heated to about 50°. A flat piece of glass may be silvered by the same liquid, which is then poured on in just sufficient quantity to cover evenly the perfectly-cleaned glass. The layer of silver so formed is very thin, and allows the passage of a violet light. It is protected from accident by a coat of paint.

433. ASSAYING OF SILVER.—The term assaying means determining the proportion of pure metal in either an alloy or an ore, but is now usually restricted to the first. Silver is alloyed with copper, and the alloy may be assayed either by a dry process—that is, one in which no liquid is employed —or by a wet process. The dry process consists in melting a small quantity of lead in a cupel, which is a little shallow cup made of compressed bone-ash and is very

FIG. 108.

porous (Fig. 108). A weighed quantity of the silver coin or jewelry to be assayed is then wrapped in a small piece of paper and placed on the surface of the melted lead, in which it is quickly dissolved. The cupel is heated in a muffle (A, Fig. 109) which fits into an opening in the side of a muffle-furnace. The muffle is open only at the exterior end, and has a slit in the arched top, so that the air is drawn through it by the draught of the furnace. The lead is oxidized by the air, and in presence of lead the copper of the alloy becomes also converted into oxide; the fused oxides are absorbed by the porous cupel, and as soon as their last traces disappear, the flashing of the bright silver surface indicates that the operation is finished. When cold, the button of pure silver is weighed.

The wet assay is an example of volumetric analysis which we must study. We know that by the addition of a solution of common salt to one of silver nitrate, silver chloride is precipitated, and, since one molecule of sodium chloride reacts with one molecule of silver nitrate, we find that 58.5 parts by weight of salt will precipitate exactly 108 parts of silver in the form of chloride.

$$\underset{(23+35.5)}{NaCl} \quad + \quad \underset{(108+14+48)}{AgNO^3} \quad = \quad \underset{(108+35.5)}{AgCl} \quad + \quad \underset{(23+14+48)}{NaNO^3}$$

By carefully adding a solution of common salt to a solution of silver nitrate, we can tell when all the silver has been converted into chloride, for no more

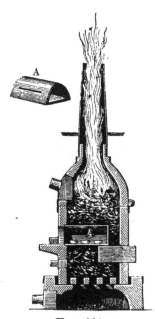

precipitate is then formed. Now, if we know how much salt we have added, we can easily calculate how much silver was present, because every 58.5 parts of salt used will represent 108 parts of silver precipitated. Let us make a solution of salt of which each litre shall precipitate ten grammes of silver. Since 108 grammes of silver require 58.5 grammes of salt, 10 grammes of silver will require $\dfrac{58.5 \times 10}{108} = 5.417$ grammes of salt. We make such a solution, and we know that every cubic centimetre of it will precipitate $\dfrac{10 \text{ grammes}}{1000} = 1$ centigramme of silver. We now dissolve in nitric acid about a gramme of our alloy of silver, accurately weighed, and then introduce our salt solution into a burette (Fig. 110), which is a glass tube having a stop-cock at the bot-

Fig. 109.   Fig. 110.

tom, and graduated so that we may measure how much of the liquid we allow to run out. Then the salt solution is slowly dropped into the solution of silver nitrate, which is agitated so that the precipitate may quickly settle, until the instant arrives when a drop produces no precipitate. We then carefully read off the exact quantity of salt solution used, and calculate the amount of silver present in the quantity of alloy analyzed, each cubic centimetre of the salt solution representing 0.01 gramme of silver.

The silver coins of the United States contain 90 per cent. of silver and 10 per cent. of copper.

434. PHOTOGRAPHY.—The chloride, bromide, and iodide of silver, being partially decomposed by the action of light, are employed in photography. An image of the object to be photographed being thrown on a glass plate coated with either of these sensitive salts, those portions on which the light falls are darkened, and metallic silver is formed; the shades or dark parts of the image remain unaffected in proportion to the intensity of the shade: then when the plate is placed in a liquid capable of dissolving the unaltered salts, a negative photograph is obtained; that is, one in which the natural lights and shades are reversed. This negative being placed over a paper sensitized by some compound alterable by light, a positive picture is obtained, for the light acts through the transparent portions of the negative. We can easily make a sensitive paper by soaking a piece of soft white paper in a solution of common salt, and, after drying it, putting it in a solution of silver nitrate in a dark room. Silver chloride is thus formed in the paper. If now we have a negative or drawing on glass, we may make a photograph; or we may copy some leaves by placing them on the paper, and, after pressing them down under a glass plate, expose the whole to the action of sunlight. In a quarter of an hour we remove the plate, and soak the paper in a solution of sodium thiosulphate (§ 106), which dissolves out the unaltered silver chloride: this is necessary, since the light would otherwise blacken the paper uniformly. After thoroughly washing the paper in water, we have an exact copy of the negative or leaves employed.

LESSON LI.

CALCIUM.—STRONTIUM.—BARIUM.

435. These three elements form a group of metals of which the corresponding compounds not only present remarkable chemical analogies, but resemble one another in many physical properties. We have already had occasion to notice, during the study of certain of their salts, that they are diatomic elements, capable of replacing two atoms of hydrogen in the acids.

The metals are obtained by decomposing their fused chlorides by a powerful electric current. They are harder than lead, and their surfaces, which are brilliant when freshly filed, rapidly tarnish in moist air. They decompose cold water, forming hydrates while hydrogen is disengaged; when heated in the air or in oxygen, they take fire and burn brilliantly.

436. **Calcium,** Ca $= 40$, is the metallic radical of lime, marble, gypsum, etc. Its density is about 1.6.

437. CALCIUM CHLORIDE, $CaCl^2$, may be made by dissolving white marble in hydrochloric acid. It is now obtained in large quantities as an accessory product in the manufacture of sodium carbonate by the ammonia process. It crystallizes in large colorless prisms containing six molecules of water of crystallization. These crystals are deliquescent; when they dissolve in water, in which they are very soluble, they produce a marked lowering of temperature. A mixture of equal weights of crystallized calcium chloride and snow or broken ice produces a temperature of —45°. When heated, the crystals melt, and at 200° four molecules of water are driven out, but the other two are retained until the temperature reaches redness. As the anhydrous calcium chloride cools, it then solidifies to a hard, white, crystalline mass; this is used for drying gases and liquids with which it undergoes no chemical reaction. Its solution in water develops considerable heat.

A saturated solution of calcium chloride boils at 179.5°. The low cost of calcium chloride obtained in the ammonia-soda process has permitted the adoption of a new and very cheap process for the extraction of sulphur from the earthy matters with which

M   23

it occurs.   The sulphur ore is immersed in a hot solution of calcium chloride of such strength that it boils at about 120°; the
sulphur then melts and runs out of the earthy matters, and may
be drawn off as it collects below the hot liquid.

438. CALCIUM OXIDE, CaO.—This substance is universally
known, and commonly called lime.   It is manufactured by decomposing limestone, which is calcium carbonate, by the action
of heat, but it is necessary that the products of combustion shall
pass through the heated mineral, for calcium carbonate is decomposed only at exceedingly high temperatures when heated in covered vessels.   Very primitive furnaces or lime-kilns are usually
employed, resembling holes in the side of a hill : above an opening at the bottom a sort of grate is arranged, and on this the coal
and limestone are thrown from the top.   The fire is then lighted,
and in about three days the kiln is burned out.   A continuous

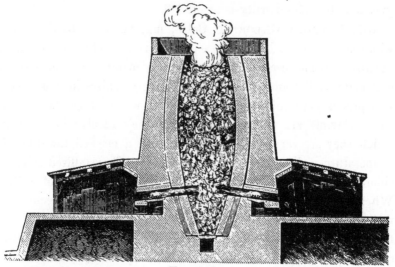

FIG. 111.

and more economical lime-kiln has an opening at the base for the
removal of the lime, and about three metres above this opening
there are others by which the flames from furnaces pass directly
into the mass of limestone.   As the lime is raked out at the bottom, the limestone descends, and more is thrown in at the top
(Fig. 111).

Lime occurs in hard, compact masses of a white or gray color : it is called *quick-lime.* It is infusible at the highest temperatures which we can produce. When exposed to the air, it absorbs moisture and carbon dioxide, cracks, increases in volume, and crumbles to a white powder, which consists of a mixture of calcium hydrate and calcium carbonate. When a mass of lime is sprinkled with water, the latter is absorbed; in a short time the lime becomes so hot that steam is given off, and, if sufficient water be used, the whole falls to a bulky powder of calcium hydrate, $Ca(OH)^2$, which is called *slaked lime.* Lime which develops much heat and increases greatly in volume by hydration is called *fat lime,* but if there be little heat produced, and the volume not greatly augmented, the lime is said to be *poor lime;* it then contains considerable quantities of magnesia, silica, and clay, derived from a poor quality of limestone. When such lime is too highly heated during the burning, calcium silicate is formed in hard, semi-fused masses, and the lime is said to be overburnt.

Milk of lime is calcium hydrate, that is, slaked lime, suspended in water. If this white, creamy liquid be allowed to settle, the clear liquid obtained is *lime-water.* It is a solution of calcium hydrate, which dissolves in about seven hundred times its weight of cold water. It is only about half as soluble in boiling water. When lime-water is heated, it becomes turbid from the separation of part of the hydrate, which again dissolves as the liquid cools.

Large quantities of lime are employed in building operations. Ordinary mortar is a mixture of slaked lime and sand, the principal object of the latter being to prevent the shrinking of the mortar as it dries. Mortar hardens because the calcium hydrate gradually absorbs carbon dioxide from the air, and the calcium carbonate formed, adhering strongly to the surfaces with which it is in contact, binds them together. It is possible that a small proportion of calcium silicate is also formed during the hardening.

Cements, of which Portland cement * is an excellent type, are

---

* Named from its resemblance to Portland stone.

made by calcining limestone with from ten to thirty per cent. of clay. Sometimes the clay exists naturally in the limestone; sometimes it is added in the form of dried river-mud. Clay is a hydrated aluminium silicate, and is rendered anhydrous by the action of heat. It is probable that at the same time a little calcium silicate and aluminate of calcium are formed. However that may be, the hard mass resulting from the calcination is pulverized, and the powder is cement, or hydraulic lime. When it is mixed with water, it sets, or hardens to a solid mass, in a very short time. It has the property of hardening under water, and is invaluable in submarine architecture. Its hardening is apparently due to the formation of a double silicate of aluminium and calcium.

439. CHLORINATED LIME, $CaCl(ClO)$.—This compound, which

FIG. 112.

is intermediate between calcium chloride, $CaCl^2$, and calcium hypochlorite, $Ca(ClO)^2$, is manufactured on an extensive scale by passing chlorine gas over well-slaked lime placed in thin layers on shelves in masonry chambers (Fig. 112), care being taken that the temperature does not become too elevated. It is largely employed as a bleaching and disinfecting agent, and owes this property to the facility with which it gives up its chlorine. It is decomposed by very dilute acids, even by the carbon dioxide of the air.

$$CaCl(ClO) \quad + \quad CO^2 \quad = \quad CaCO^3 \quad + \quad Cl^2$$

When thrown into water, it yields a solution containing calcium hypochlorite and calcium chloride.

$$2CaCl(ClO) \quad = \quad CaCl^2 \quad + \quad Ca(ClO)^2$$

Chlorinated lime.   Calcium hypochlorite.

When it is heated, or when its solution is boiled, it is converted into calcium chloride and calcium chlorate.

$$6CaCl(ClO) \quad = \quad 5CaCl^2 \quad + \quad Ca(ClO^3)^2$$

Chlorinated lime.   Calcium chlorate.

440. Tests for Calcium.—Solutions of calcium salts are not affected by hydrogen sulphide. In solutions which are not very dilute, sulphuric acid and the soluble sulphates produce a white precipitate of calcium sulphate. Solution of oxalic acid to which a few drops of ammonia have been added, yields a white precipitate of calcium oxalate, even in the most dilute calcium solutions. The salts of calcium communicate a reddish-yellow color to flame, and the calcium spectrum is quite characteristic. (See frontispiece.)

441. Strontium, Sr = 87.5.—The principal strontium minerals are the sulphate, called *celestine*, on account of the blue color of many specimens, and the carbonate, called *strontianite*. The first, being the more abundant, serves for the preparation of the strontium salts: it is powdered, and intimately mixed with charcoal, and the mixture heated to bright redness in a covered crucible. Carbon monoxide is then disengaged, while the sulphate is reduced to the sulphide, SrS. The gray mass containing this sulphide is then treated with the acid corresponding to the desired salt, which separates in crystals when the solution is evaporated.

442. Strontium Chloride, SrCl², is made by dissolving the sulphide in hydrochloric acid and evaporating the filtered solution. It crystallizes in deliquescent needles containing three molecules of water of crystallization for one molecule of the salt. It is very soluble in water, and slightly soluble in alcohol, the latter solution burning with a red flame.

443. STRONTIUM MONOXIDE, SrO, is prepared by strongly calcining strontium nitrate obtained by dissolving the sulphide in nitric acid. It is an infusible, gray, porous mass: when exposed to the air, it absorbs moisture and carbon dioxide. By the action of water, it is converted into strontium hydrate, $Sr(OH)^2$, which is soluble in about fifty times its weight of cold or two and a half times its weight of boiling water, and may.be obtained in crystals with eight molecules of water of crystallization.

444. STRONTIUM DIOXIDE, SrO^2, is formed by the action of oxygen on the monoxide at a dull red heat.

445. TESTS FOR STRONTIUM.—Solutions of the ordinary salts of strontium are colorless; they are not precipitated by hydrogen sulphide. Sodium carbonate produces a voluminous white precipitate of strontium carbonate. Sulphuric acid precipitates strontium sulphate in solutions which are,not too dilute. Oxalic acid and ammonia produce a white precipitate of strontium oxalate. Flame is colored red by strontium compounds.

·446. **Barium**, $Ba = 137$.—Barium occurs in nature in *heavy-spar*, which is the sulphate, and *witherite*, which is the carbonate. Its salts may be prepared by dissolving the native carbonate in the corresponding acid, or from the sulphate, which must first be reduced to sulphide. The finely-powdered sulphate is made into a paste with rosin and linseed oil, and the mixture is shaped into little balls which are calcined in a covered crucible.

447. BARIUM CHLORIDE, $BaCl^2$, is obtained when the sulphide is dissolved in hydrochloric acid, and the filtered solution sufficiently concentrated. Its crystals contain two molecules of water. They are soluble in rather more than twice their weight of cold water, in much less boiling water, and also slightly soluble in alcohol. Barium chloride is the reagent generally used for the detection of sulphuric acid.

448. BARIUM MONOXIDE, BaO, often called baryta, is prepared, like strontium monoxide, by calcining the nitrate. It forms a gray, porous mass, which absorbs moisture and carbon dioxide from the

air. If a fragment of this substance be sprinkled with a few drops of water, barium hydrate is formed with such energy that the mass sometimes becomes red hot. ·

449. BARIUM HYDRATE, BaOH, is made by dissolving the oxide in boiling water, which dissolves about one-tenth its weight. When the liquid cools, the greater part of the hydrate is deposited in colorless crystals which contain $Ba(OH)^2 + 8H^2O$. These crystals are soluble in water, and, under the name baryta-water, their solution is used for the precipitation of carbon dioxide as insoluble barium carbonate, or for the precipitation of sulphuric acid.

450. BARIUM DIOXIDE, BaO^2.—At a dull red heat, barium monoxide will absorb oxygen, and become converted into the dioxide, which is made by passing oxygen over the monoxide heated in a porcelain tube or in a crucible. Barium dioxide is a grayish-white substance, which, when thrown into water, crumbles to a white hydrate. It loses one atom of oxygen at a bright red heat, and the monoxide remains. By the action of strong sulphuric acid, barium dioxide is converted into barium sulphate, while ozone is disengaged. With hydrochloric acid, the hydrated dioxide yields barium chloride and hydrogen dioxide.

451. TESTS FOR BARIUM.—Hydrogen sulphide occasions no precipitate in solutions of barium salts. Sodium carbonate throws down white barium carbonate. Sulphuric acid precipitates insoluble barium sulphate, even in exceedingly dilute solutions, and the precipitate is insoluble in nitric acid, either cold or boiling. Barium salts communicate a green color to flame.

The barium salts are very poisonous.

452. The nitrates of barium and strontium are employed in pyrotechny, for they impart to fireworks the characteristic flame colors of the metals.

A red fire may be made by mixing 30 parts of potassium chlorate, 17 parts of sulphur, 2 of charcoal, and 45 of strontium nitrate. The materials must be pulverized separately, and may be mixed by repeated passing through a sieve. A green fire may be made by similarly mixing 33 parts of potassium chlorate, 10 of sulphur, 5 of charcoal, and 52 of barium nitrate. If it be desired that the fires shall produce little or no smoke, the following formulæ may be used; the ammonium picrate may be made by adding ammonia-water to a concentrated alcoholic solution of picric acid, until the liquid has an ammoniacal odor, and then collecting and carefully drying the precipitate.

	Ammonium picrate.	Ferrous picrate.	Strontium nitrate.	Barium nitrate.
Yellow	50	50	...	...
Green	48	...	...	52
Red	54	...	46	...

The stars for rockets and Roman candles are made by moistening the colored fires and forming them into small balls; these are dried and introduced into the tube, from which they are projected by a small charge of gunpowder.

LESSON LII.

LEAD. $Pb = 207$.

453. In many of its chemical relations, lead resembles calcium, strontium, and barium, and it might be classed in the same group of metals; but in a number of its compounds it acts as a tetratomic element. It forms a dioxide, PbO^2, and a tetrachloride, $PbCl^4$. In the dioxides of strontium and barium, it is not probable that the atoms of these metals are tetratomic: it appears rather that the two atoms of oxygen are related to each other, while each is also related to the atom of metal. In lead dioxide the lead atom is tetratomic.

454. The principal lead ores are *galena*, which is lead sulphide, and *cerusite*, which is the carbonate. The reduction of the latter mineral is an exceedingly simple process: it is heated with char-coal; the reduced lead collects on the hearth of the furnace, and is drawn off as it accumulates.

Galena may be reduced by heating it with scrap iron: iron sulphide and lead are formed, and, the lead being the heavier, the iron sulphide floats on the surface and is drawn off as slag. The more usual process, known as the reaction process, consists in heating the galena on the hearth of a reverberatory furnace (Fig. 112) provided with openings (D) for the admission of air. Part of the lead sulphide is so converted into oxide, and another portion into sulphate. When this reaction has sufficiently advanced, the openings of the furnace are closed, and the heat is increased. Under these circumstances the unaltered sulphide reacts with both oxide and sulphate, metallic lead being formed, while sulphur dioxide is disengaged.

$$PbS + 2PbO = 3Pb + SO^2$$
$$PbS + PbSO^4 = 2Pb + 2SO^2$$

Sometimes charcoal powder is added after the air-openings are closed, in order to aid in the reduction of the oxide and sulphate.

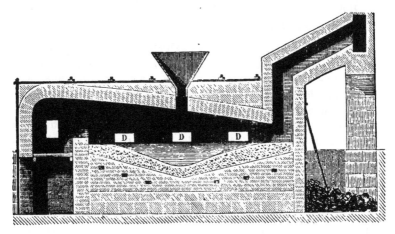

Fig. 112.

Lead is a bluish-white metal, having a' brilliant lustre, which soon tarnishes by exposure to air. It is soft, and can be scratched by the finger-nail: it is quite malleable, but has so little tenacity that it cannot readily be drawn into wire. Its density is about 11.36 : it melts at about 330°. It may be crystallized by allowing a crucible full of the molten metal to cool until a crust forms on its surface, piercing the crust, and pouring out the still liquid interior. The interior of the crucible is then found to be lined with octahedral crystals. Molten lead absorbs oxygen from the air, and its surface becomes covered with a film of lead oxide, PbO.

Lead is dissolved by boiling hydrochloric acid, lead chloride being formed while hydrogen is disengaged. It is scarcely affected by dilute sulphuric acid, but the strong acid dissolves it by the aid of heat, sulphur dioxide being given off. Nitric acid converts it into lead nitrate, and disengages red vapors.

Pure water containing dissolved air and carbon dioxide dissolves a small quantity of lead in the form of hydrate and carbonate, and for this reason lead is an unsafe metal for lining rain-water cisterns intended for storing drinking-water. Most spring- and river-waters contain small quantities of sulphates: when such water flows through lead pipes, the surface of the metal becomes quickly covered with a film of insoluble lead sulphate, which protects the pipe from further action, and the water from being poisoned by the introduction of lead compounds.

Lead, and all its soluble compounds, as well as such as may be rendered soluble by the juices of the stomach, are poisonous, and the poisonous effects are cumulative. Workmen employed in the manufacture of white lead, red lead, and other lead compounds, frequently suffer from chronic lead-poisoning, as do also painters and color-grinders. Small quantities of lead are then accumulated in the system, and cause peculiar disorders, among which *lead colic* is the most common: one of the characteristic symptoms of lead-poisoning is a peculiar blue line around the borders of the gums. The workmen in lead-works usually drink small quantities of an exceedingly dilute sulphuric acid, by which the lead in the system is converted into the insoluble and innocuous sulphate. In cases of chronic lead-poisoning, the administration of potassium iodide removes the metal from the tissues by the formation of lead iodide, which is soluble in solutions of potassium iodide, and can consequently be eliminated by the excretory organs.

Metallic lead is used in the form of sheets for roofing and lining tanks; it is manufactured into lead pipe; type-metal, which is 80 per cent. lead and 20 per cent. antimony; pewter, which contains between eighty and ninety per cent. tin, the remainder being lead; and plumbers' solder, an alloy of lead and tin. Enormous quantities of lead are employed for the manufacture of shot, which is made by allowing the molten metal to run through a sieve, and the drops to fall from a height into water. In common qualities of tin plate, a large proportion of the coating is lead instead of pure tin.

455. LEAD CHLORIDE, $PbCl^2$, is prepared by boiling lead oxide in hydrochloric acid, and is precipitated when hydrochloric acid or a soluble chloride is added to the solution of a lead salt. It is a white solid, only slightly soluble in cold water, but dissolving in thirty-three times its weight of boiling water: when the hot solution cools, the chloride separates in brilliant anhydrous needles. It is employed in the manufacture of several yellow colors, which are oxychlorides of lead, or mixtures of the chloride and oxide.

456. LEAD IODIDE, PbI^2, is deposited as a yellow precipitate when potassium iodide is added to the solution of a lead salt. It

is almost insoluble in cold water, but dissolves in a little less than two hundred times its weight of boiling water, from which it separates on cooling in beautiful golden-yellow scales.

457. LEAD MONOXIDE, PbO.—This body is produced by the direct oxidation of melted lead by the air. It is an accessory product in the cupellation of lead for the extraction of silver (§ 426). It is known in commerce by the names *massicot* and *litharge*: massicot is a yellow, amorphous powder; by fusing this powder and pulverizing the resulting mass, litharge is obtained as reddish-yellow, crystalline scales. Lead monoxide is slightly soluble in water, and will restore the blue color to reddened litmus. It melts at a red heat, but cannot be melted in vessels of glass, porcelain, or clay, because it combines with silica and forms a very fusible silicate, so destroying the vessel. It is readily reduced by charcoal and by hydrogen. It is used in the manufacture of the salts of lead: when it is boiled with linseed oil, the latter acquires the property of quickly drying or hardening when exposed to the air.

When the solution of a lead salt is treated with an alkaline hydrate, lead hydrate, $Pb(OH)^2$, is thrown down as a white precipitate, soluble in an excess of the alkaline hydrate.

458. LEAD DIOXIDE, PbO^2, is obtained by treating red lead with nitric acid. Red lead is a combination of the monoxide and dioxide, and the nitric acid dissolves out the monoxide, forming lead nitrate, which is soluble and can be washed out, while the dioxide remains as a brown powder. It is not soluble in water, and by the action of heat is decomposed into lead monoxide and oxygen. It is a very energetic oxidizing agent: a little sulphur may be ignited by rubbing it in a mortar with some lead dioxide. It absorbs sulphur dioxide, forming lead sulphate; with hydrochloric acid it forms lead chloride and water, while chlorine is disengaged.

$$PbO^2 + 4HCl = PbCl^2 + 2H^2O + Cl^2$$

459. RED LEAD, $(PbO)^2PbO^2$.—This body is prepared by heating massicot to 300° in furnaces so arranged that it is freely exposed to a current of air; oxygen is then absorbed, and a beautiful red powder, called *minium*, or red lead, is formed. It is

plumboso-plumbic oxide, but the proportions of the monoxide and dioxide which it contains are not constant, though usually responding to the formula given. When heated, its color darkens; at a red heat it loses part of its oxygen and is converted into the monoxide. Red lead is employed as a pigment, and in the manufacture of flint-glass, of which the brilliancy and refractive power are due to silicate of lead. Mixed into a paste with linseed oil, it forms an excellent cement.

460. LEAD SULPHIDE, PbS.—This compound is the mineral *galena*, which is found in cubical crystals of a bluish-gray color and metallic appearance. Its density is 7.58; it is much harder than lead, and rather brittle. It melts when heated to redness, and in contact with air is then oxidized to sulphide and sulphate. It is converted into lead chloride by boiling with hydrochloric acid, hydrogen sulphide being disengaged. Boiling nitric acid converts it into lead sulphate.

461. TESTS FOR LEAD.—With the exception of the nitrate and acetate, none of the more common lead salts are very soluble. Those which are soluble have a sweet and somewhat astringent taste. Hydrogen sulphide forms in them a black precipitate of lead sulphide: potassium and sodium hydrates and ammonia produce white precipitates, which are soluble in an excess of either of the first two reagents. Sulphuric acid yields a white precipitate even in the most dilute solutions: hydrochloric acid throws down white lead chloride, unless the solution be too dilute; this precipitate is dissolved by boiling, and, on cooling again, separates in crystals. Potassium chromate precipitates yellow lead chromate, which is soluble in the alkaline hydrates.

If a salt of lead be mixed with sodium carbonate, and heated on a piece of charcoal in the inner flame of a blow-pipe, a small bead of metallic lead is obtained, and the softness of the bead indicates the nature of the metal.

LESSON LIII.

MAGNESIUM.—ZINC.—CADMIUM.

462. These three metals form a natural group, to which belongs also a fourth, glucinum, of which the silicate constitutes part of the mineral beryl and the green precious-stone emerald. They are diatomic metals.

463. **Magnesium, Mg = 24.**—This element occurs in nature as carbonate in the mineral *magnesite*, as sulphate in *kieserite*, and as silicate in *serpentine* and *soapstone.* The metal is obtained by heating its chloride with sodium in an iron crucible, a mixture of common salt and calcium fluoride being added as a flux. The sodium is converted into sodium chloride, and the magnesium separates in little globules diffused through the molten mixture, which is constantly stirred. When perfectly cold, the mass is broken up, and the globules of magnesium are removed and heated to redness in a small charcoal vessel in a current of hydrogen. Pure magnesium volatilizes, and condenses in the cooler part of the apparatus.

Magnesium is a bluish-white metal; its surface, which is not very brilliant, soon tarnishes in the air. Its density is about 1.75. It is both ductile and malleable, and is ordinarily rolled into ribbon or drawn into wire. It slowly decomposes water at ordinary temperatures, and more rapidly at the temperature of boiling. It melts at 500°, and if exposed to the air takes fire and burns with great brilliancy. The light of burning magnesium is very bright, and lamps are constructed in which the ribbon is gradually supplied by clock-work. Such lamps are employed in photographing the interior of caves and other dark localities. The product of the combustion is magnesium oxide, MgO.

464. MAGNESIUM CHLORIDE, $MgCl^2$.—When magnesium or its oxide or carbonate is dissolved in hydrochloric acid, and the solution is concentrated, crystals of magnesium chloride, with six molecules of water of crystallization, are obtained. These crystals

cannot be rendered anhydrous, and their solution cannot be evaporated to dryness, for they decompose into hydrochloric acid and magnesium oxide.

$$MgCl^2 \quad + \quad H^2O \quad = \quad MgO \quad + \quad 2HCl$$

Anhydrous magnesium chloride is prepared by dissolving the oxide or carbonate in hydrochloric acid, and adding two molecules of ammonium chloride for every atom of magnesium. This solution may be evaporated to dryness, and leaves an anhydrous double chloride of magnesium and ammonium. The double salt is heated in a clay crucible until all of the ammonium chloride is driven off, while the magnesium chloride remains in a state of fusion ; on cooling, it solidifies to a pearly-white mass. In this form it is used for the manufacture of magnesium. It is very soluble in water, but from the solution only the hydrated crystals can be obtained.

465. MAGNESIUM OXIDE, MgO.—This is the calcined magnesia of the pharmacies. It is made by calcining magnesium carbonate, or the mixture of hydrate and carbonate commonly called white magnesia. It is a tasteless white powder, infusible even at very elevated temperatures. It is insoluble in water, but combines with that liquid, forming magnesium hydrate, $Mg(OH)^2$, a substance which restores the blue color to reddened litmus. This same hydrate is precipitated when an alkaline hydrate is added to the solution of a magnesium salt.

466. TESTS FOR MAGNESIUM.—Neither hydrogen sulphide nor ammonium sulphide occasions any precipitate in magnesium solutions. Sodium carbonate throws down a white, flocculent precipitate of the hydrated carbonate, which when dried in the air constitutes white magnesia. Potassium and sodium hydrates yield white precipitates of the hydrate, as does also ammonia unless the solution be acid or contain ammonium chloride. Sodium phosphate with a few drops of ammonia produces a white, crystalline precipitate of ammonio-magnesium phosphate, $Mg(NH^4)PO^4$.

467. **Zinc**, $Zn = 65.2$.—The ores from which zinc is obtained are the carbonate, which is called *calamine*, and the sulphide, called *blende*. These minerals are broken up and roasted in fur-

naces much resembling lime-kilns. At the temperature of the roasting, which is a dull red heat, the calamine loses carbon dioxide and the water which it usually contains, and is converted into zinc oxide: the sulphide is also oxidized by roasting, sulphur dioxide being disengaged. The zinc oxide so obtained is mixed with charcoal and heated for about twenty-four hours to a high temperature in clay or iron vessels: carbon monoxide is disengaged, while the zinc volatilizes and is condensed in suitable apparatus. Various processes of distillation are employed: we need only consider the two which are generally used. In the Belgian process, the mixture of zinc oxide and charcoal is introduced into clay tubes, closed at one end, and inserted in an inclined position in the walls of the furnace; to the open and exterior end of each tube is adapted a bulged pipe, in which the zinc vapor condenses and the metal collects. In order that no air may enter the tubes and oxidize the zinc, a sheet-iron noz-zle, having a hole for the exit of the gases, is passed over the extremity of this condenser (Fig. 114). The tubes are usually about a metre in length, and twenty centimetres in inte-

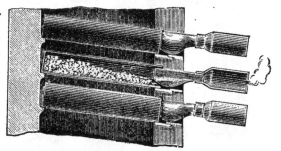

FIG. 114.

rior diameter. A large number of them are placed in parallel rows in the same furnace: when all the zinc has distilled, the receivers containing it are removed, and a fresh charge of roasted ore and charcoal is introduced into the tubes.

In the Silesian process, the retorts are arched, and very similar in form to those employed in the manufacture of illuminating gas from coal.

468. At present, the furnace used in the reduction of zinc by both the Belgian and Silesian methods is that known as the Siemens regenerative furnace, which effects a great saving of fuel. In this arrangement, the coal is fed gradually to the grate of a peculiar fire-box, called the generator, and the admission of air is there so regulated that as much carbon monoxide as possible may

be produced by an imperfect combustion; in addition, the ashes below the grate are kept moist, and the steam passing into the fire reacts with the hot carbon, producing hydrogen and carbon monoxide ($\S$ 232); the highly-heated gas is led through a chamber filled with fire-bricks, which become very hot; by a system of dampers, the gases are then directed through another similar chamber, while air is admitted to that which has been heated; the heated air from the one, and the heated gas from the other, are then brought in contact where it is desired that the greatest temperature shall be produced by the perfect combustion of the gases. The heat of the waste products of combustion is applied to heating other fire-brick chambers, which will afterwards serve for

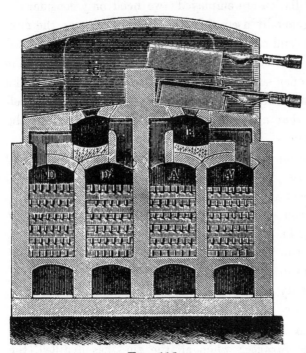

FIG. 115.

the admission of air, as these regenerators, as they are called, are cooled by the entering air. Figure 115 represents the fire-brick chambers of a Siemens furnace applied to the Silesian zinc process. The two chambers on each side serve alternately, one for the entrance of air, and one for the gas from the generator, while the other two serve for the exit of the products of combustion. The heated air and gas from A and A′ come in contact in the space B, and the flames play through openings in the floor above which are the clay retorts. The heated products of combustion pass over the retorts in another similar chamber, C, and from above downwards through other fire-brick chambers, D and D′. The dampers allow the direction of the current of gas and air to be reversed from A A′ to D D′ as often as necessary, and in practice it is so changed about once every hour.

Zinc must usually be purified before it is sent into commerce, and the most harmful impurity is lead, for it impairs the malleability of the zinc. The lead is separated in great part by melting the zinc in moulds which are slightly inclined and have a cavity at the lower end: in this the greater part of the lead collects by reason of its greater density, and may be broken from the cooled ingot. Commercial zinc usually contains small quantities of iron, copper, lead, cadmium, and sometimes arsenic. Sheet zinc is the purest.

Zinc is a bluish-white metal, capable of taking a high lustre. Its density varies from 6.86, that of the cast metal, to 7, that of the rolled. Pure zinc may be hammered into sheets, or drawn into wire at ordinary temperatures, but commercial zinc must be rolled at about 150°. It again becomes brittle at 200°, and may readily be pulverized in a mortar heated to that temperature. It melts at 410°, and distils at about 1000°. It is unaltered by dry air, but in moist air its surface becomes dull from the formation of a film of hydrated carbonate, which protects the metal from further action.

When it is heated to redness in the air, it takes fire and burns with a bluish flame, giving off clouds of white zinc oxide, ZnO. Fine zinc shavings may be lighted by a match, and burn brilliantly in the air. If some zinc be heated to redness in a ladle or crucible, and pieces of potassium nitrate be thrown in, the oxygen of the decomposing nitre energetically oxidizes the metal.

Zinc is dissolved by hydrochloric, sulphuric, and nitric acids, and by boiling solutions of potassium and sodium hydrates. In the latter case, hydrogen is disengaged and an alkaline zincate is formed, a compound in which zinc oxide appears to act as an acid radical. We have already studied the action of the acids on zinc.

Zinc is employed in the manufacture of galvanized iron, which is made by dipping carefully cleaned iron objects into melted zinc; brass, which is an alloy of copper and zinc; the plates of voltaic batteries; and for the preparation of zinc white, which is zinc oxide.

469. ZINC CHLORIDE, $ZnCl^2$, may be formed by the direct union of zinc and chlorine, a union which takes place brilliantly when fine zinc shavings are thrown into a jar of chlorine. It is prepared by dissolving zinc in hydrochloric acid. It forms deli-

quescent crystals containing one molecule of water of crystallization, which is expelled by heat, and the anhydrous salt fuses at
250°. The latter is very deliquescent, and is an energetic deby-
drating agent. It is employed as a caustic in surgery. Zinc
chloride is very soluble in water, and its solution, to which a little
free hydrochloric acid and some ammonium chloride have been
added, is an excellent soldering liquid, for moistening the surface
of iron, zinc, copper, and brass articles before soldering.

470. ZINC OXIDE, ZnO, is prepared on a large scale by heating
zinc in large muffles in which its vapor may come freely in contact with air. The product is stirred up with water; the heavier
particles of unaltered zinc sink to the bottom, while the zinc oxide
remains suspended in the creamy liquid which is rapidly poured
off and allowed to settle. The separation of fine powders by this
method is called *elutriation*.

Zinc oxide is a white powder, insoluble in water. It is employed as a substitute for white lead in painting localities exposed
to hydrogen sulphide, which would blacken a lead pigment.

When an alkaline hydrate is added to the solution of a zinc
salt, zinc hydrate, $Zn(OH)^2$, is thrown down as a white precipitate.

$$ZnSO^4 \;+\; 2KOH \;=\; K^2SO^4 \;+\; Zn(OH)^2$$

This precipitate is soluble in an excess of the alkaline hydrate.

471. ZINC SULPHIDE, ZnS.—This compound is found native
as zinc blende, a mineral usually having a more or less intense
brown color, due to the presence of a certain proportion of iron.
When ammonium sulphide is added to the perfectly neutral solution of a zinc salt, a white precipitate of hydrated zinc sulphide is
formed.

472. TESTS FOR ZINC.—Neutral solutions of zinc salts are
precipitated white by hydrogen sulphide; the precipitate is not
formed if free mineral acid be present. Ammonium sulphide
produces a characteristic white precipitate of zinc sulphide. The
alkaline hydrates and ammonia-water yield white zinc hydrate,
soluble in an excess of the reagent. Potassium ferrocyanide
throws down a white precipitate of zinc ferrocyanide. The salts
of zinc are poisonous.

473. **Cadmium,** $Cd = 112$.—This metal occurs associated with zinc in both blende and calamine. It is reduced with the zinc, and, being more volatile than the latter, distils during the early part of the operation. During the first few hours of the reduction of many zinc ores, a brown powder, called cadmies, collects in the receivers attached to the retorts. This dust contains a large proportion of cadmium oxide, and when distilled with charcoal powder yields an alloy of zinc and cadmium. The latter metal is purified by dissolving the alloy in dilute sulphuric acid, precipitating cadmium sulphide by passing hydrogen sulphide through the acid liquid, dissolving the sulphide in hydrochloric acid, and adding ammonium carbonate. Cadmium carbonate is precipitated; this is collected, dried, and roasted, and the cadmium oxide obtained is distilled with charcoal powder.

Cadmium has a white color and a brilliant lustre, which soon becomes dull in moist air. Its density is 8.60. It melts at 320° and boils at 860°. Hydrochloric and sulphuric acids dissolve it rapidly, disengaging hydrogen.

474. CADMIUM IODIDE, CdI^2, is made by digesting cadmium filings and iodine in water. On evaporating the solution, beautiful transparent and colorless hexagonal prisms of cadmium iodide are deposited. It is used in photography.

475. CADMIUM OXIDE, CdO, is obtained as a yellowish-brown powder by roasting either cadmium nitrate or cadmium carbonate. It is reduced by hydrogen and carbon at lower temperatures than those required for the corresponding reductions of zinc oxide.

476. CADMIUM SULPHIDE, CdS.—This compound is found in nature in brilliant yellow, hexagonal prisms. It is precipitated as an amorphous yellow powder by the action of hydrogen sulphide on solutions of cadmium salts. It is employed as a pigment by artists.

477. TESTS FOR CADMIUM.—Potassium and sodium hydrates and ammonia-water give white precipitates of cadmium hydrate; only that formed by ammonia is soluble in an excess of the reagent. Hydrogen sulphide throws down a characteristic yellow precipitate of cadmium sulphide, even in acid solutions. Potassium ferrocyanide gives a yellowish-white precipitate of cadmium ferrocyanide.

LESSON LIV.

COPPER.

478. Large deposits of metallic copper exist on the shores of Lake Superior, the metal being sometimes found in crystals, sometimes in irregular and grotesque masses. The more common

copper ores are cuprous sulphide, called *chalkosine*, and *copper pyrites*, a compound of cuprous sulphide and ferrous sulphide. This metal is also found as cuprous oxide, cupric oxide, and cupric carbonate.

Pure copper ores—those containing only the oxide, carbonate, or sulphide of copper, and very little of other metals—are easily reduced: the sulphide is first converted into oxide by roasting, and the ores are then heated with charcoal in a somewhat conical furnace. The reduction of copper pyrites is more difficult, especially if, as is often the case, this mineral be mixed with the sulphides of antimony, arsenic, zinc, etc. If such ore contains a large proportion of copper, it may be worked by a dry process; but if only a small percentage of copper is present, a method of solution is adopted. In the dry process, the ore is first roasted by being fed from hoppers on to the hearth of a reverberatory furnace (Fig. 116), where it is swept by the flame of a fire. Part of

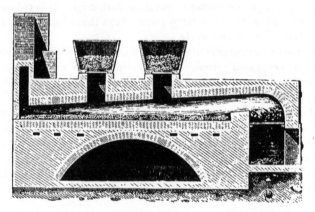

Fig. 116.

the sulphur is so converted into sulphurous oxide, which may be used for the manufacture of sulphuric acid, while the iron and copper of the pyrites are partially converted into oxide and sulphate. A quantity of sand and silicate of iron from a subsequent stage of the operation is then added, and the mass is transferred either to rotating cylindrical furnaces or to reverberatory furnaces with deep hearths, where it can be strongly heated. The unaltered ferrous sulphide remaining in the roasted mass then reacts with the cupric oxide formed, and the result is cuprous sulphide and ferrous oxide. The latter unites with the sand, forming ferrous silicate, which is very fusible, and is drawn off as slag, while a tolerably pure form of cuprous sulphide collects on the hearth of the furnace. This product, which is called copper matt, is broken up, and repeatedly roasted until nearly all the sulphur is expelled, and a considerable proportion of the copper is reduced to the metallic state; the more oxidizable foreign metals present become oxidized, and, on the addition of silicious matters, are converted into fusible silicates by an increased temperature. The *black copper* so obtained contains from 90 to 94 per cent. of copper, the remainder being lead, iron, sulphur, arsenic, etc. It is refined by melting it on the hearth of a reverberatory furnace, where the foreign metals are completely oxidized and removed as slags, while the copper collects in a cylindrical cavity made in the hearth of the furnace. It is solidified by throwing water on its surface, and the circular masses removed are called copper rosettes. They are brittle, for they consist partly of cupric oxide. This is reduced by melting the rosettes under a layer of charcoal powder, and stirring the molten metal with poles of green wood; the combustible gases formed by the action of the high temperature on the wood completely deoxidize the copper, and the cold metal is red and soft.

In the wet process, used for poor copper ores, the latter are roasted with from one to two or three per cent. of common salt in peculiar rotating furnaces. Hydrochloric acid and sulphurous and sulphuric oxides are disengaged; by passage through a tall column filled with coke, over which cold water trickles, these gases are absorbed for subsequent use in another stage of the operation. The roasted mass contains cupric chloride and cupric sulphate: it is pulverized and washed with the water in which the gases from the furnace have been condensed. The copper salts then pass into solution, while ferric oxide remains. The latter is dried and heated with charcoal powder, by which it is converted into a very spongy metallic iron. This iron is introduced into the copper solution: iron salts are formed, and all the copper is quickly deposited in the metallic state, together with any lead, arsenic, antimony, and silver which might be present. The precipitated metal is collected and melted into a mass.

Sometimes copper contains enough silver to pay for its extraction. This may be accomplished by
melting the metal with lead and cast-
ing the alloy into disks: these are
stood on edge over a gutter in a furnace in which they may be heated gradually (Fig. 117). The lead melts first,
and runs out, carrying with it all of
the silver. The mass of copper is
again melted and freed from the last
traces of lead, while the lead which has removed the silver is
submitted to cupellation.

Fig. 117.

Pure copper is obtained by placing scrap iron in solution of
pure cupric sulphate, and thoroughly washing the precipitated
copper, which is called cement copper. In certain copper localities, streams of running water contain sufficient dissolved copper
sulphate, produced by the natural oxidation of copper pyrites
on the lands which supply the streams, to pay the cost of the
extraction of the metal. This is effected by placing scrap iron in
troughs through which the stream is made to flow; the copper
then deposits on the iron.

Copper has a red color and a brilliant lustre. Its density is
about 8.9. It is exceedingly ductile, malleable, and tenacious.
It melts at about 1100°, and may be crystallized either by fusion
or by electrolysis of solutions of its salts. In contact with the
skin, it produces a very unpleasant odor.

Copper is unaltered by cold dry air, but by moist air it is
gradually converted into a hydrocarbonate, which appears in green
spots on the surface of the metal. This is the substance ordinarily
called verdigris (see § 333).

At a temperature about redness, copper combines directly with
oxygen, forming either cupric oxide, CuO, or cuprous oxide,
Cu^2O, according to the access of air. When copper acetate is
strongly heated in a hard glass tube, it is entirely decomposed,
and a residue of finely-divided copper is obtained. If this be
turned out and heated at one point by a lighted match, a black

spot appears and rapidly spreads over the entire mass, which is so converted into cupric oxide.

We have already studied the action of sulphuric and nitric acids on copper. Hydrochloric acid attacks it only when boiling, and then but slowly, evolving hydrogen and forming cuprous chloride, Cu^2Cl^2.

Ammonia in presence of oxygen exerts a curious action on copper. We introduce some copper clippings and a little ammonia into a bottle, which we tightly cork and then agitate for a few minutes. The liquid becomes blue, and if we invert the bottle and open it with its mouth under water, the latter will rise in the bottle, showing that part of the air has been absorbed. It is the oxygen which is absorbed, and the blue liquid contains copper nitrite and ammoniacal cupric oxide, both of which are soluble in ammonia. This liquid is capable of dissolving cotton, linen, paper, and other forms of cellulose.

Copper is used for the manufacture of boilers, stills, condensing apparatus, and other utensils for the laboratory, manufactory, and kitchen. In sheets, it serves for sheathing ships, and sometimes for roofing. It constitutes part of many alloys, among which are brass, containing from 65 to 90 per cent. copper, the remainder being zinc; a large proportion of copper gives a red color to the brass; these metals are melted together in crucibles, the zinc being added after the copper is fused. Bronze contains from 93 to 95 per cent. of copper, the remainder being tin, with sometimes 1 per cent. of zinc. Gun metal is about 91 per cent. copper and 9 per cent. tin. Bell-metal and the very white speculum metal contain respectively 78 and 67 per cent. of copper, the remainder being tin. German silver is an alloy of copper, zinc, and nickel. The United States cents contain 95 per cent. of copper, 2.5 per cent. of zinc, and 2.5 per cent. of tin.

479. Copper forms two series of compounds. It is a diatomic element, and in the cuprous compounds two atoms of copper form a diatomic couple, Cu–Cu, which replaces two atoms of hydrogen in the acids. In the cupric compounds, a single diatomic atom of copper replaces two atoms of hydrogen.

480. Cuprous Chloride, Cu^2Cl^2, may be made by boiling a solution of cupric chloride with copper, or by boiling copper with hydrochloric acid and adding a little nitric acid from time to time;

in the latter case, cupric chloride is formed, and is at once reduced by the metallic copper present. On adding water to the brown liquid so obtained, cuprous chloride is thrown down as a white crystalline precipitate. It is insoluble in water, but dissolves in ammonia, forming a colorless solution which absorbs oxygen and becomes blue on exposure to air. It also dissolves in hydrochloric acid, and both of these solutions are capable of absorbing a large volume of carbon monoxide.

A hydrated compound of cuprous chloride and cupric oxide constitutes the beautiful green mineral *atacamite*.

481. CUPRIC CHLORIDE, $CuCl^2$, is obtained when cupric oxide is boiled in hydrochloric acid. When the green solution is sufficiently concentrated, it deposits beautiful bluish-green crystals of cupric chloride, with two molecules of water of crystallization.

482. CUPROUS OXIDE, Cu^2O.—This substance occurs in nature, sometimes in amorphous masses, sometimes in red, regular octahedra. It may be made by boiling glucose with a solution of cupric acetate, and is then thrown down as a bright-red crystalline precipitate. If heated in contact with air, it is converted into cupric oxide. It is used for imparting a red color to glass, being added to the mixture of sand and sodium carbonate.

483. CUPRIC OXIDE, CuO.—When cupric nitrate is strongly heated, it yields a fine black powder of cupric oxide. This compound is usually prepared by heating metallic copper to redness in vessels through which air is blown or drawn. The copper then absorbs oxygen, and is converted into hard and compact cupric oxide. This substance is reduced by both hydrogen and charcoal at temperatures below redness, water or carbon dioxide being formed. It communicates a green color to glass, and is used for that purpose. In the laboratory it is of great value in the analysis of carbon compounds.

When potassium or sodium hydrate is added to the solution of a cupric salt, cupric hydrate, $Cu(OH)^2$, is formed as a pale-blue precipitate. When the liquid containing this hydrate is boiled, the precipitate turns black, for it is converted into cupric oxide and water, even when surrounded by liquid.

484. Cuprous Sulphide, Cu^2S.—This is the mineral *chalkosine*. It may be obtained as a black, brittle, crystalline mass by fusing together sulphur and copper, or by burning copper in vapor of sulphur.

485. Cupric Sulphide, CuS, is thrown down as a brownish-black precipitate by the action of hydrogen sulphide on cupric solutions. When heated, it loses sulphur, and is converted into cuprous sulphide.

486. Carbonates of Copper.—When a solution of cuprous sulphate is treated with sodium carbonate, carbon dioxide is disengaged, and a bluish-green precipitate is thrown down; when washed with warm water, its color becomes green; it is a compound of cupric hydrate and cupric carbonate, containing $CuCO^3.Cu(OH)^2$. The beautiful green mineral *malachite*, which when polished displays veins of variegated tints, is a compound having the same composition. *Azurite*, a mineral found in fine blue crystals, is a compound of two molecules of cupric carbonate with one of cupric hydrate, $Cu(OH)^2.2CuCO^3$.

487. Tests for Copper.—The salts of copper have either blue or green colors. Both hydrogen sulphide and ammonium sulphide throw down brownish-black precipitates. The alkaline hydrates precipitate pale-blue cupric hydrate, insoluble in an excess of the reagent. Ammonia also produces a pale-blue precipitate, but this dissolves when an excess of ammonia is added, yielding a magnificent blue solution of an ammonio-cupric salt.

Potassium ferrocyanide produces a mahogany-brown precipitate of cupric ferrocyanide, and the test is exceedingly delicate and characteristic. A clean piece of iron, as a needle or knife-blade, dipped in a cupric solution, quickly becomes covered with a red layer of metallic copper: this test is conclusive.

LESSON LV.

MERCURY. Hg = 200.

488. Mercury is found in small quantity in the metallic state, but its principal ore is the sulphide, which constitutes the mineral *cinnabar*.   It is especially abundant in Spain and on the Pacific slope.

The reduction of cinnabar is a simple operation: it is broken up and roasted in a current of air, the sulphur being expelled as sulphur dioxide, while mercury distils.   Very little improvement has been effected in the furnaces during hundreds of years; the mercury vapor is sometimes condensed by being passed through a long series of clay pipes, sometimes by being directed through a number of chambers containing a layer of water, by which the gases are cooled (Fig. 118). The mercury is then filtered through

Fig. 118.

closely-woven canvas, and is usually transported in iron bottles, each bottle holding about sixty pounds.

Mercury is liquid at ordinary temperatures: it freezes at —40°, and boils at 350°.   Its density at 0° is about 13.6.

The density of mercury vapor compared with that of hydrogen is 200 : its atomic weight is also 200, as is shown by the vapor-densities of its volatile compounds. Then if equal volumes of gases contain equal numbers of molecules, and if the molecule of hydrogen contain two atoms, the molecule of mercury vapor must consist of a single atom. We believe that the molecule and atom of cadmium are identical also, and for a similar reason.

Mercury is unaffected by the air at ordinary temperatures, but at 300° it absorbs oxygen and is converted into red mercuric oxide. It combines directly, and in the cold, with chlorine, bromine, and iodine, and with sulphur by the aid of a gentle heat, or if the sulphur be finely divided. Mercury is not dissolved by hydrochloric acid : boiling sulphuric acid converts it into mercuric sulphate, sulphur dioxide being disengaged. Nitric acid dissolves it, emitting red vapors, and forming mercurous nitrate if the reaction take place in the cold, or mercuric nitrate if the acid be boiling.

Mercury is used for filling thermometers, barometers, and pressure-gauges; for silvering ordinary mirrors, which are coated with tin foil amalgamated with mercury; for the extraction of silver and gold from their ores; and for the preparation of various amalgams.

The mercury of commerce is rarely pure; it contains small quantities of lead, copper, tin, and sometimes bismuth. Its approximate purity may be determined by allowing a few drops to fall on a clean piece of paper or porcelain; pure mercury will then break up into small globules which are perfectly round, and move about freely when the surface on which they rest is inclined, but mercury containing other metals forms globules that are drawn out to a tail, and that do not move so readily. The surface of pure mercury is perfectly brilliant, but when impure the metal has a tarnished appearance. It may be purified by shaking it in a bottle with sulphuric acid and a solution of potassium dichromate : the impurities are thus oxidized, and may be washed away.

Like copper, mercury is diatomic, and forms two series of compounds,—mercurous compounds, in which two atoms form a diatomic couple, and mercuric compounds, in which two atoms of hydrogen are replaced by a single diatomic mercury atom.

489. Mercurous Chloride, Hg^2Cl^2.—This compound is the well-known medicine calomel. It is made by subliming a mixture of mercurous sulphate and common salt.

$$Hg^2SO^4 \quad + \quad 2NaCl \quad = \quad Na^2SO^4 \quad + \quad Hg^2Cl^2$$
Mercurous sulphate.   Mercurous chloride.

The calomel then condenses in appropriate receivers, in dense crystalline masses. It is usually resublimed, and its vapors passed into jars or chambers filled with steam, where it condenses in an impalpable powder. Calomel is precipitated when hydrochloric acid is added to the solution of a mercurous salt.

In masses, calomel occurs in dense, fibrous, crystalline, and translucent fragments, colorless when recently prepared, but becoming gray or yellowish by the action of light which partially decomposes this compound into mercuric chloride and mercury. Its density is about 7.2. It is insoluble in water; when calomel is agitated with water and the liquid filtered, no turbidity should be produced in the filtrate by the addition of sodium carbonate solution. If mercurous chloride be heated with a solution of sodium chloride, it is converted into mercuric chloride, while metallic mercury is deposited as a gray powder.

490. Mercuric Chloride, $HgCl^2$, is the poisonous compound corrosive sublimate. It is prepared by subliming a mixture of common salt and mercuric sulphate, sodium sulphate being formed at the same time.

$$HgSO^4 \quad + \quad 2NaCl \quad = \quad Na^2SO^4 \quad + \quad HgCl^2$$
Mercuric sulphate.   Mercuric chloride.

It is also formed by the direct combination of chlorine and mercury.

It forms dense, white or colorless, crystalline masses, having a density of 6.5. It melts at 265°, and boils at about 295°. It is soluble in nineteen times its weight of cold water, and in much less boiling water, from which it separates in anhydrous crystals on cooling. It is exceedingly poisonous, and its antidote is white of egg, for it forms an insoluble compound with albumen.

491. Mercurous Iodide, Hg^2I^2, is obtained as a green powder by rubbing together in a mortar 100 parts of mercury and 63.5

parts of iodine with a few drops of alcohol. By the action of light or heat, it is decomposed into mercuric iodide and metallic mercury.

492. MERCURIC IODIDE, HgI^2.—This beautiful compound is prepared by mixing potassium iodide with four-fifths its weight of mercuric chloride, both in aqueous solution, and thoroughly washing the precipitate.

$$HgCl^2 + 2KI = HgI^2 + 2KCl$$

If either substance be employed in excess, the precipitate will be redissolved.

So obtained, mercuric iodide forms a dark-red powder, which is almost insoluble in water, but dissolves slightly in boiling alcohol, and on cooling separates in red, octahedral crystals.

Mercuric iodide presents a curious case of dimorphism. If a little of the red powder be cautiously heated on a sheet of white paper on which it is spread out, the red color changes to yellow; the yellow particles are rhombic prisms, and if they be rubbed with a glass rod or any hard body, they will reassume the red color and their first crystalline form, the octahedron. Mercuric iodide melts to a dark-yellow liquid, and volatilizes, condensing in the yellow crystals.

With potassium iodide, mercuric iodide forms a soluble compound, which may be obtained by dissolving the mercuric iodide in solution of potassium iodide. The colorless liquid is called Nessler's reagent, and is used in the laboratory as a test for ammonia, and compound ammonias, with which it forms a brownish cloud or a dense precipitate, according to the proportion of ammonia present.

493. MERCUROUS OXIDE, Hg^2O.—This substance is obtained as a black powder by digesting mercurous chloride in a solution of potassium hydrate. A temperature of 100°, or the prolonged action of light, decomposes it into mercuric oxide and mercury.

494. MERCURIC OXIDE, HgO, has long been known under the name red precipitate. It may be made either by decomposing mercuric nitrate by heat until the whole is converted into a red powder and no more red vapors are disengaged, or by adding po-

25*

tassium hydrate to a solution of mercuric chloride and thoroughly washing the precipitate. Prepared in the first manner, it forms a red, crystalline powder; obtained by precipitation, it is yellow and amorphous, but becomes red when heated.

Mercuric oxide is insoluble in water: when it is heated, its color darkens, and at a temperature of about 400° it is decomposed into metallic mercury and oxygen. It is an energetic oxidizing agent. In presence of water, it converts chlorine into hypochlorous acid, and when dry and quite cold, into hypochlorous oxide. If a mixture of a little mercuric oxide and sulphur be heated in a test-tube, it explodes.

495. MERCURIC SULPHIDE, HgS.—This is the mineral *cinnabar*, which is found in hard dense masses, and in transparent red crystals. It is manufactured by grinding together the required proportions of mercury and sulphur, and subliming the resulting black mass. It then forms a dark-red, crystalline solid, having a density of 8.12. When strongly heated out of contact with air, it volatilizes without melting. When heated in the air, it takes fire, and burns with a blue flame, mercury vapor and sulphur dioxide being disengaged.

The fine scarlet pigment *vermilion* is very finely divided mercuric sulphide, made by grinding for a long time in a mortar a mixture of 300 parts of mercury and 114 parts of flowers of sulphur: 75 parts of potassium hydrate, dissolved in 400 parts of water, are then added, and the grinding is continued, the mortar being kept at a temperature of about 45°. When the powder has assumed the desired shade, it is quickly washed with hot water, and dried.

496. TESTS FOR MERCURY.—Very few of the mercurous salts are soluble: in their solutions, hydrochloric acid produces a white precipitate of mercurous chloride; hydrogen sulphide and potassium and sodium hydrates and ammonia produce black precipitates.

With mercuric salts, hydrogen sulphide and ammonium sulphide give black precipitates; potassium hydrate throws down yellow mercuric oxide. If a piece of bright copper be dipped into the slightly acid solution of either a mercurous or a mercuric salt, metallic mercury is quickly deposited on the copper, whose surface becomes white and brilliant after a little friction.

When heated with lime or sodium carbonate in a small glass tube, all compounds of mercury yield a sublimate of metallic mercury, which condenses in the cooler part of the tube in microscopic globules. On throwing a fragment of iodine into the still warm tube, the glóbules are changed into yellow or red mercuric iodide.

LESSON LVI.

BISMUTH AND GOLD.

These two metals are triatomic: they form chlorides whose molecules contain one atom of metal and three atoms of chlorine. They form trioxides, containing two atoms of metal and three of oxygen, and gold forms an oxide, Au^2O.

497. Bismuth, $Bi = 210$.—Bismuth is found in the metallic state disseminated in quartz. It is separated from the earthy materials, which are called the *gangue*, by heating the mineral in iron tubes which are closed at one end, and arranged in an in-clined position in a furnace beyond which the lower and open end projects. The bismuth then melts and runs out of the tubes. The bismuth thus obtained is never pure, but contains small quantities of other metals, and sometimes traces of arsenic and sulphur. In order to purify it, it is pulverized and mixed with a little potassium nitrate: the mixture is heated to redness in clay crucibles; the impurities, which are more easily oxidized than the bismuth, are thus oxidized, and any arsenic present is con-verted into potassium arsenate.

Bismuth is a crystalline, brittle, yellowish-white metal. Its density is 9.8. It melts at 264°. By allowing a crucible full of the molten metal to cool until a crust forms on the surface, and then pouring out the liquid interior through a hole made in the crust, fine crystals of bismuth may be obtained. These crystals become superficially oxidized, and the thin film of oxide imparts to them all the colors of the rainbow. Bismuth is unaffected by cold air, but at a red heat it is burned to bismuth oxide. It dissolves in nitric acid, forming bismuth nitrate, while red vapors are disengaged.

In addition to its use for the preparation of the bismuth compounds, this metal is employed chiefly for the manufacture of certain alloys. Britannia metal contains about one per cent. of bismuth. The fusing points of the bismuth alloys are much lower than that of bismuth. A mixture known as Wood's alloy or fusible metal consists of one or two parts of cadmium, two parts of tin, four of lead, and seven or eight of bismuth. It melts between 66° and 71°, according to its composition. Another alloy, known as Arcet's fusible metal, is made by melting together eight parts of bismuth, five of lead, and three of tin. It melts at 94.5°.

Bismuth much resembles antimony in many of its chemical relations ; but we class it among the metals, because it is capable of replacing the hydrogen of oxygen acids, so forming well-defined salts.

498. Bismuth Chloride, $BiCl^3$.—This compound results from the direct union of chlorine and bismuth. When powdered bismuth is sprinkled into chlorine, it burns brilliantly, forming the chloride. This substance is prepared by passing dry chlorine over melted bismuth in a retort so arranged that the chloride may collect in a receiver as it distils. It then forms a crystalline deliquescent mass, which is quite soft at ordinary temperatures, being very fusible. It is soluble in hydrochloric water, but is decomposed by water, hydrochloric acid being formed, while a white powder of bismuth oxychloride, BiOCl, is thrown down.

$$2BiCl^3 \quad + \quad 2H^2O \quad = \quad 4HCl \quad + \quad 2BiOCl$$

Bismuth oxychloride constitutes the cosmetic known as pearl-white.

499. Bismuth Oxide, Bi^2O^3, is obtained as a yellow powder when bismuth nitrate is strongly heated. It melts at a red heat, and on cooling solidifies to a glassy, yellow mass. It forms a very fusible silicate, and therefore cannot be melted in clay crucibles. Bismuth hydrate, probably $Bi(OH)^3$, is thrown down as a white powder when bismuth subnitrate is treated with potassium hydrate or ammonia-water.

There are other oxides of bismuth, corresponding exactly to the oxides of nitrogen.

500. Bismuth Nitrate, $Bi(NO^3)^3$.—When bismuth is boiled with nitric acid, and the solution is concentrated, large, colorless,

deliquescent crystals of bismuth nitrate with three molecules of water of crystallization are deposited. Since bismuth is triatomic, one atom of bismuth will replace the hydrogen in three molecules of nitric acid, and combine with the three groups NO^3. The crystals of bismuth nitrate are very soluble in water containing free nitric acid; but if the solution be diluted with a large volume of water, a pulverulent white precipitate is thrown down. This contains $(BiO)NO^3$, or $BiNO^4$, and is employed in medicine under the name subnitrate of bismuth. A larger quantity may be obtained by adding very dilute ammonia to the liquid.

501. TESTS FOR BISMUTH.—When solutions of the bismuth salts are largely diluted with water, white precipitates of sub-salts are thrown down. Hydrogen sulphide and ammonium sulphide occasion brown precipitates of bismuth sulphide. The alkaline carbonates and hydrates yield white precipitates, insoluble in an excess of the reagent.

When a bismuth salt is heated with sodium carbonate in the inner flame of a blow-pipe, a brittle bead of metallic bismuth is obtained.

502. **Gold**, $Au = 197$.—Gold is found in the metallic state, sometimes in masses called nuggets, but more usually in small particles disseminated through quartz rock, or the sand produced by the disintegration of the rock. It is sometimes associated with silver, copper, lead, and tellurium. The gold is extracted from gold-bearing sand by washing the latter in a stream of running water in troughs called cradles. By reason of its great density, the gold then sinks to the bottom, while the lighter sand is carried on with the water. The gold may then sometimes be removed at once; sometimes it is in such small particles that it must be amalgamated with mercury, as will presently be described. Quartz rock containing gold is crushed by powerful machinery, and the greater part of the earthy matter is removed by washing in vessels containing mercury, which forms an amalgam with the gold. Fig. 119 represents an apparatus which is sometimes employed for grinding together the mercury and crushed rock. It consists of inclined iron basins, each containing two cast-iron balls: the rock

and mercury being introduced into these vessels, a motion of rotation is communicated by machinery, and by the friction of the

FIG. 119.

balls the rock is reduced to an impalpable powder, which is carried off by a current of water flowing through the basins, while the gold amalgamates with the mercury. The amalgam is compressed in chamois-skin or canvas bags, through which the excess of mercury is forced, and the solid amalgam remaining in the bags is then heated in iron retorts. The mercury distils, while the gold remains.

On the Pacific slope large quantities of gold are obtained by hydraulic mining, which is conducted by throwing streams of water with great force against the soft and disintegrated rocks containing the gold. The stream of water, flowing from these rocks, and carrying with it the gold and mud, and even large stones, is conducted through sluices into settling-troughs containing mercury, which removes the gold.

As extracted from its native rocks or sand, gold is rarely pure. It usually contains more or less silver; this may be removed by boiling the metal in nitric acid, which does not affect the gold, while it converts the silver into silver nitrate. However, if only a small proportion of silver be present, that metal is protected by the gold, and it is necessary to melt the alloy with a larger proportion of silver before boiling it with nitric acid. The gold then remains as a spongy mass. Pure gold may also be obtained by adding ferrous sulphate or oxalic acid to a solution of gold chloride; in this case the gold is thrown down as a dark-brown, dull powder, capable of assuming its natural high lustre by burnishing.

The color of gold varies from greenish yellow to a red almost as decided as that of copper. Light which has been successively reflected from ten surfaces of gold is scarlet. Gold is very soft, and the most malleable and ductile of the metals. Its density is 19.5; it melts at about 1200°, and at a higher temperature emits a green vapor. A thin gold-leaf, carefully spread out between two plates of glass, allows the passage of a faint green light.

Gold is not oxidized by air, either moist or dry, or at any temperature. It is not affected by boiling with nitric, sulphuric, or hydrochloric acids. Nitro-hydrochloric acid dissolves it, disengaging red vapors, and forming a yellow solution of gold trichloride; the nitro-hydrochloric acid employed for dissolving gold is a mixture of nitric acid with four times its weight of hydrochloric acid. Gold is also attacked by selenic acid, H^2SeO^4, by a hot mixture of iodic and sulphuric acids, and by a boiling mixture of concentrated nitric and sulphuric acids; from the latter solution the gold is again deposited in the metallic state by the addition of water. Gold dissolves readily in chlorine-water, in bromine, and combines directly with iodine under the influence of light.

Gold forms two series of compounds,—aurous compounds, in which the metal appears to be monatomic, and auric compounds, in which it is triatomic.

503. AURIC CHLORIDE, $AuCl^3$.—When the solution of gold in nitro-hydrochloric acid is evaporated, auric chloride is deposited as a dark-red crystalline mass, which is very deliquescent. It is very soluble in water and in ether. Its strong solutions are orange brown, but the dilute solution is pure yellow. It produces a violet stain on the skin; it is decomposed by the action of heat, and more slowly by light, and is reduced by many substances, among which are phosphorus, phosphorous and sulphurous acids, oxalic acid, and ferrous sulphate. A stick of phosphorus immersed in an ethereal solution of auric chloride becomes quickly coated with a film of gold. The metal is deposited as a brown powder

when either ferrous sulphate or oxalic acid is added to a solution of auric chloride.

When a solution containing a mixture of stannous and stannic chlorides is added to auric chloride, a flocculent, purple precipitate of uncertain composition, but containing gold, tin, oxygen, and hydrogen, is thrown down. This precipitate is known as purple of Cassius, and is employed in painting on glass and porcelain.

When auric chloride is heated to 230°, chlorine is disengaged, and an insoluble yellow powder of aurous chloride, $AuCl$, remains.

There are two oxides of gold,—aurous oxide, Au^2O, and auric oxide, Au^2O^3. The first is basic, the second forms aurates with the metals. When caustic alkalies are fused with gold in contact with air, alkaline aurates are formed.

504. ASSAYING OF GOLD.—Gold coin, jewelry, etc., are generally alloyed with silver, and sometimes with copper. A weighed quantity of the metal to be assayed is first melted with about three times its weight of silver, and the resulting button is cupelled in a bone-ash cupel ($\S$ 433). The copper and any other base metals present are so converted into oxides, which are absorbed by the cupel, and a button containing only gold and silver is obtained. This is hammered out into a thin sheet, which is twisted up and boiled in nitric acid; the silver is dissolved, while the gold remains as a spongy mass, which is washed, heated to redness, and then weighed.

The gold coin of the United States contains 90 per cent. of gold, the remainder being copper.

505. GILDING.—Silver and copper objects may be gilded by rubbing over them an amalgam of gold with eight times its weight of mercury. They are then heated under a chimney so arranged that the poisonous mercury vapor may be entirely carried off. The dull gilded surface is then rendered brilliant by burnishing. A thin film of gold is deposited on copper objects when they are dipped into a hot solution of auric chloride with sodium carbonate and sodium phosphate.

Gilding is best accomplished by connecting the objects to be gilded with the zinc pole of a voltaic battery, and immersing them in a solution obtained by boiling auric chloride with potassium cyanide. The positive pole of the battery is connected with a plate of gold immersed in the same liquid.

The rare elements *indium* and *thallium*, both of which were discovered by the aid of the spectroscope, are related to gold and bismuth in the general constitution of their compounds. Traces of indium exist in most zinc blendes, while small quantities of thallium occur in certain iron pyrites, and the metal is obtained from the dust which collects in the flues of sulphuric acid works when these pyrites are burned for the production of sulphur dioxide.

LESSON LVII.

ALUMINIUM. Al $= 27.5$.

506. Although this is one of the most abundant elements, it is found only in combination, and the preparation of the metal is a matter of much difficulty. It is made by throwing a mixture of sodium and double chloride of aluminium and sodium on the red-hot hearth of a reverberatory furnace from which the air is excluded. Aluminium chloride may be used instead of the double chloride mentioned, but the latter is more easily prepared, and is not so readily altered by contact with air as the former. Some cryolite is added to the mixture to serve as a flux. The reaction results in the formation of sodium chloride and aluminium.

$$Al^2Cl^6, 2NaCl \quad + \quad 3Na^2 \quad = \quad 8NaCl \quad + \quad Al^2$$
Aluminium and sodium double chloride. \hspace{4cm} Aluminium.

The globules of aluminium are then fused together, and cast into ingots.

Aluminium is a bluish-white metal, capable of being highly polished. It is very ductile and malleable, and also very sonorous. It is a good conductor of heat and electricity. Its density is 2.56; it is therefore as light as glass and porcelain. It melts at about 750°. It is unaltered by the air at ordinary temperatures, but when melted absorbs oxygen and is converted into aluminium oxide. It is hardly affected by either nitric or sulphuric acid, but dissolves readily in hydrochloric acid, disengaging hydrogen and forming aluminium chloride. It is also dissolved by boiling solutions of the alkaline hydrates, hydrogen being set free, while alkaline aluminates are formed.

The great tenacity of aluminium, and its lightness and unchangeableness in the air, render it an exceedingly valuable metal, but unfortunately the high cost of its production has prevented its employment for many purposes to which it is admirably adapted.

Aluminium is a tetratomic metal, but the molecules of its compounds contain two atoms of aluminium, which together form a hexatomic couple, capable of combining with six atoms of chlorine, three of oxygen, etc.

507. ALUMINIUM CHLORIDE, Al^2Cl^6.—When aluminium or its hydrate is dissolved in hydrochloric acid, a solution of aluminium chloride is obtained, but this solution cannot be evaporated to dryness without decomposing into aluminium oxide and hydrochloric acid.

$$Al^2Cl^6 \quad + \quad 3H^2O \quad = \quad Al^2O^3 \quad + \quad 6HCl$$

Solid aluminium chloride is formed by passing chlorine gas over a red-hot mixture of aluminium oxide and charcoal, which has been made into small balls with a little oil, and then calcined in a crucible. These balls are put in a clay tube or retort, which is heated to bright redness, and dry chlorine is then passed through (Fig. 120). Carbon monoxide and aluminium chloride are

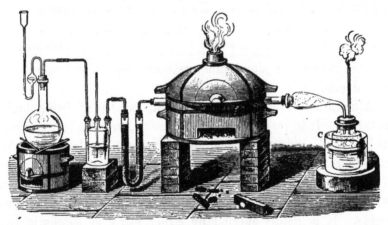

FIG. 120.

formed, and the latter, being volatile, must be condensed in a bottle surrounded with cold water.

$$Al^2O^3 \quad + \quad 3C \quad + \quad 3Cl^2 \quad = \quad 3CO \quad + \quad Al^2Cl^6$$
Aluminium oxide.   Aluminium chloride.

Aluminium chloride is a white or pale-yellow crystalline compound, which melts at a gentle heat, and volatilizes at a temperature slightly above 100°. When thrown into water, it dissolves and combines with the liquid, forming a hydrate, which cannot be dried without decomposition. It slowly absorbs moisture from the air, giving off hydrochloric acid while aluminium oxide is formed. The double chloride of aluminium and sodium which is used in the preparation of aluminium, melts at about 200°; it is made when common salt is added to the mixture of aluminium oxide and charcoal used in the preparation of aluminium chloride.

508. ALUMINIUM OXIDE, Al^2O^3.—This compound is commonly called *alumina*. It is found native in corundum, ruby, sapphire, topaz, and emery: the black color of the latter is due to the presence of oxide of iron. Aluminium oxide may be obtained in the laboratory by heating the aluminium hydrate which is thrown down as a gelatinous white precipitate when ammonia is added to a solution of alum. It then forms a white powder, which is infusible except at the temperature of the oxyhydrogen blow-pipe flame; it is not reducible by either hydrogen or charcoal. The crystallized varieties of alumina are used as gems: ruby is red, sapphire is blue, and topaz is yellow. By reason of their hardness, corundum and emery are of great value in grinding and polishing glass, steel, and metals.

Aluminium hydrate, $Al^2(OH)^6$, forms a bulky gelatinous precipitate when ammonia-water or an alkaline hydrate or carbonate is added to a solution of alum or any salt of aluminium.

509. ALUMINIUM SULPHATE, $Al^2(SO^4)^3$.—Clay is a silicate of aluminium, usually colored yellow by the presence of a little iron. By boiling with strong sulphuric acid, clay is decomposed, a solution of aluminium sulphate being formed. This body is made from clay as free as possible from iron; when its solution is evaporated, a white crystalline mass is obtained, and by special precautions the salt may be crystallized in small pearly scales or needles containing eight molecules of water of crystallization. It is soluble in twice its weight of cold water, and is used as a mordant in dyeing, for it may be decomposed in the fibres of the tissues to be dyed, and the fine particles of aluminium oxide deposited firmly fix the color in the fabric. For this purpose it is usually first converted into aluminium acetate by the addition of calcium acetate.

When aluminium sulphate is heated, it first loses it., water of crystallization, and then gives off sulphur trioxide, leaving a residue of aluminium oxide.

$$Al^2(SO^4)^3 = 3SO^3 + Al^2O^3$$

510. ALUMS.—To a cold saturated solution of aluminium sulphate, we add a cold saturated solution of potassium sulphate, and stir the mixture. A crystalline deposit forms. The two salts have combined to form a double salt, which is called an alum. It crystallizes with twenty-four molecules of water of crystallization, and its formula is $Al^2(SO^4)^3.K^2SO^4 + 24H^2O$. By the substitution of sodium sulphate or ammonium sulphate for the potassium sulphate in the preceding experiment, sodium alum or ammonium alum will be formed. The compositions of these substances are precisely analogous to that of the potassium alum, and they crystallize in the same form, which is the regular octahedron.

$Al^2(SO^4)^3.Na^2SO^4 + 24H^2O$ Sodium alum.

$Al^2(SO^4)^3.(NH^4)^2SO^4 + 24H^2O$ Ammonium alum, ordinary alum.

Potassium alum is soluble in about thirty times its weight of cold water, and in less than one-third its weight of boiling water. It forms voluminous, transparent crystals when the hot saturated solution is allowed to cool. When heated, it melts in its water of crystallization, which is afterwards driven off; the salt increases enormously in volume, and the anhydrous alum then forms a white, porous mass. Alum may be obtained crystallized in cubes by adding a very small quantity of potassium carbonate or hydrate to its hot solution and allowing it to cool.

Sodium alum is very soluble in cold water, and is not employed in the arts.

Ammonium alum is the compound ordinarily called alum. Its solubility is about the same as that of potassium alum. When it is strongly heated, it leaves a residue of pure alumina.

Other metals whose oxides are analogous in constitution to aluminium oxide, form alums having compositions and general properties like those of ordinary alum. These alums are isomorphous. Although their colors be different, they may be mixed in the same crystal, and the form of the latter will remain unchanged. Thus, chromium alum is red: when one of its red octahedral crystals is immersed in a saturated solution of potassium alum and the water is allowed to evaporate, the octahedron will grow larger, and the red chromium alum will be surrounded by the colorless potassium alum. The compositions of three of these alums are shown by the following formulæ:

Iron alum, $Fe^2(SO^4)^3.K^2SO^4 + 24H^2O$

Manganese alum, $Mn^2(SO^4)^3.K^2SO^4 + 24H^2O$

Chromium alum, $Cr^2(SO^4)^3.K^2SO^4 + 24H^2O$

511. CLAY AND POTTERY.—*Feldspar, albite,* and *labradorite* are double silicates of aluminium and potassium, sodium and calcium, respectively. *Granite* and *mica* are also double silicates of aluminium with the alkaline silicates or calcium silicate. The disintegration of these rocks by the action of air and frost results in the formation of clays, and the nature of a clay will depend on that of the rock from which it is derived. The purest clay is a hydrated silicate of aluminium known as *kaolin,* or porcelain clay. It contains $Al^2O^3.2SiO^2.2H^2O$. Clays which form a coherent mass when mixed with water, and which when calcined become very hard without being fused, are called plastic clays, and are used for the manufacture of bricks, fire-brick, pottery, etc. *Fuller's earth* is a kind of clay of which the paste is not strongly coherent: it is used in scouring and fulling cloths. *Marls* are mixtures of clay and chalk, generally of a greenish color, and often found in large deposits: they are used as fertilizers for sandy soils.

Porcelain is made from a mixture of the finest kaolin with a little finely-powdered sand and feldspar, which are added to prevent the mass from shrinking and to render the ware translucent by undergoing partial fusion. The greatest care is exercised that the materials, which are made into a paste with water, may be intimately mixed; after the articles have been fashioned from the perfectly homogeneous paste, they are baked at a dull red heat, and, after cooling, are removed from the furnace. They are then dipped into water holding in suspension a mixture of kaolin and quartz in an impalpable powder. This powder fills the pores on the surface, and when the articles are again baked the mixture fuses and forms a transparent glaze, while the whole mass becomes partially vitrified.

Stoneware is manufactured from a kaolin which is not sufficiently pure for porcelain-making. It is baked at one operation, and when the temperature of the oven is very high a little common salt is thrown on the incandescent objects; by the action of the hydrogen compounds in the flame, hydrochloric acid is formed, while the sodium forms a double silicate with aluminium on the surface of the ware. This silicate, being quite fusible, melts and spreads out on the surface of the ware, forming an even glaze.

Articles of faïence are made from a still more common clay mixed with finely-powdered quartz, and, after being rendered coherent by a preliminary baking, are coated with a mixture of powdered quartz, potassium carbonate, and lead oxide. This mixture fuses to a transparent varnish when the articles are baked a second time, and various colors are obtained by the addition of certain metallic oxides. Oxide of tin renders the glaze white and opaque.

The glazing of pottery intended for culinary purposes should contain no lead, as lead silicate is attacked by dilute vegetable acids, and a lead salt is sometimes so formed in articles of food.

512. TESTS FOR ALUMINIUM.—Solutions of aluminium salts usually have an acid reaction. The alkaline hydrates and ammonia produce gelatinous white precipitates of aluminium hydrate, soluble in acids and in the alkaline hydrates. The same precipitate is thrown down by the alkaline carbonates and by ammonium sulphide, carbon dioxide being liberated by the former and hydro-gen sulphide by the latter.

$$Al^2(SO^4)^3 + 3(NH^4)^2S + 3H^2O = Al^2(OH)^6 + 3(NH^4)^2SO^4 + 3H^2S$$

When an aluminium salt or aluminium oxide is strongly heated in a blow-pipe flame, and the resulting white mass is moistened with a drop of cobalt nitrate solution and again heated, it becomes sky-blue, without fusion.

513. Closely related to aluminium by chemical analogies are a number of rare metals, of which four have been obtained in the metallic state. They are lanthanum, cerium, didymium, and gallium. The first three exist as silicates in the mineral *cerite* and in other rare minerals: gallium occurs in exceedingly minute quantities in certain zinc blendes.

Scandium, samarium, holmium, erbium, thulium, and yttrium are elements which have not been obtained in the metallic state; but their oxides have been isolated in small quantities, and a few of their salts have been studied. Each of these elements is distinctly characterized by a peculiar spectrum, which is an unquestionable indication of the individuality of the element. All these elements appear to form sesquioxides, and their chlorides, which have been prepared, contain two atoms of metal and six of-chlorine, corresponding in composition to aluminium chloride.

LESSON LVIII.

IRON AND ITS METALLURGY.

514. Iron is found in the metallic state in meteoric stones, which are occasionally drawn to the earth during its passage through space.

The more important iron ores are ferroso-ferric oxide, Fe^3O^4, called *magnetic iron ore*, or black oxide of iron; ferric oxide, Fe^2O^3, called *red hematite;* ferric hydrate of various composition, known as *brown hematite, öolite, goethite, bog iron ore*, etc.; ferrous carbonate, $FeCO^3$, called *spathic iron;* and iron disulphide, FeS^2, which is *iron pyrites.* Excepting in comparatively rare cases, these ores are mixed with a greater or less proportion of clay, silicious matters, etc. When the ore contains sulphur, it is first roasted, and the sulphur is burned into sulphur dioxide, while the iron remains as oxide. The sulphur dioxide produced by the roasting of pyrites is often employed in the manufacture of sulphuric acid.

The oxide of iron, either the natural ore or produced by the roasting, is reduced by being heated with charcoal. A rather primitive method, but one which for ages furnished all of the iron employed, and which is still used for the reduction of very rich ores, is known as the Catalan method, the name being derived from the Spanish province in which the process is still carried on. It consists in piling the ore and charcoal in two heaps, side by side, on burning charcoal contained in the hearth of a furnace where the combustion is sustained by a blast of air from the tuyère of a bellows (Fig. 121). The reduced iron collects on the hearth in a spongy mass, which is removed and directly submitted to the operation of forging. The silicious matters of the ore combine with a portion of ferrous oxide produced during the operation, and form a very fusible slag, consisting of ferrous silicate.

515. The *blast-furnace process* for the reduction of iron is applicable to all iron ores, and the fuel employed is either charcoal, coke, or anthracite. The blast-furnace is a tall column of considerable height, sometimes almost cylindrical, but more often constructed in the form of a double frustum of cones placed base to base (Fig. 122). It is lined with infusible fire-brick; the hearth is flat, and inclines very slightly towards the front, which is so arranged that the molten iron may be drawn off at the bottom through a hole which is kept closed with a clay plug, and the slag may be removed as it accumulates and floats on the surface of the liquid iron. During the reduction, the bottom of the furnace is closed, and a blast of air is injected through tuyère pipes (T) by powerful blowing engines. Coal, ore, and limestone are continually supplied in alternate layers at the open top of the furnace, and in the interval between the introduction of these materials the top is closed by a conical cap or dome, which can be readily moved by suitable machinery. At the beginning of the operation the furnace is heated by a supply of combustibles only, and, when the temperature has been sufficiently raised, the ore and limestone are gradually alternated with the introduction of coal, until the furnace is completely filled. By the combustion of the coal immediately above the tuyères, carbon dioxide is produced, but as this comes in contact with the highly-heated coal it is reduced to carbon monoxide: as the latter gas rises through the mixture, it reduces the oxide of iron, and the metallic iron formed is disseminated in small particles through the mass of reduced ore. As the materials descend in the furnace, the silicious matters of the ore, and the lime resulting from the

Fig. 121.

action of the heat on the limestone, unite to form a very fusible slag of calcium and aluminium silicate, while the particles of iron are agglomerated together, and, together with the slag, flow to the hearth of the furnace. The gases produced during the operation contain a large proportion of carbon monoxide; they are carried

Fig. 122.

off by pipes inserted near the top of the furnace, and their combustion furnishes heat for the boilers which supply steam to the blowing engines and to furnaces containing long series of pipes, through which the air from the engines is forced before it enters the tuyères. The blast is heated as highly as possible, for by the use of hot air a very great saving is effected in the quantity of fuel required.

When a sufficient quantity of iron has accumulated on the hearth of the furnace, the blowing engines are slowed or stopped, and by picking out the clay plug the molten iron is caused to flow into semi-cylindrical channels in sand on the floor of the casting-room. Blows from a sledge-hammer detach the bars of iron so formed from that in the channel from which the moulds are filled, and they constitute pig-iron.

This iron contains carbon and a small proportion of silicon and traces of sulphur and phosphorus, which it has removed from the materials in the blast-furnace. These impurities are in great part removed by melting the iron in puddling-furnaces, where the carbon and silicon are oxidized either by air, or better by oxygen derived from pure magnetic iron ore, or from scales of black oxide of iron obtained in another operation. During the process, the molten iron is vigorously stirred until it is converted into a spongy mass, which is then removed and placed under a steam-hammer, by the blows of which all the ferrous silicate and black oxide of iron formed during the process of puddling are squeezed out, while a *bloom* of soft iron remains. The ferruginous scoriæ or ashes obtained in this operation are used in refining a new quantity of cast-iron.

516. The soft iron of commerce, called forged or bar iron, is not perfectly pure. It contains traces of carbon, silicon, sulphur, phosphorus, nitrogen, and sometimes other elements. Pure iron may be obtained by passing hydrogen over pure ferric chloride heated to bright redness in a porcelain tube. Hydrochloric acid is disengaged, and the iron remains as an almost infusible, spongy mass. By passing dry hydrogen over ferric oxide heated to dull redness in a glass bulb (Fig. 123), metallic iron is obtained as a dull black powder, in which form it becomes oxidized with great readiness. If a lighted match be applied to a single point in recently-prepared iron reduced by hydrogen, the whole mass quickly takes fire and burns into ferric oxide. By reducing the ferric oxide at a temperature below redness, a powder of iron may be obtained which will even take fire spontaneously on contact with the air.

The composition and general properties of cast-iron vary greatly, for while cast-iron always contains silicon and carbon, these elements do not appear to be chemically combined with the iron. The proportion of carbon varies from 2 to 5.5 per cent. When cast-iron containing a large proportion of carbon is rapidly cooled, it becomes hard and brittle, and its fracture is coarse, crystalline, and very white. It is called *white iron*. When, however, such iron is allowed to cool slowly, a considerable quantity of the

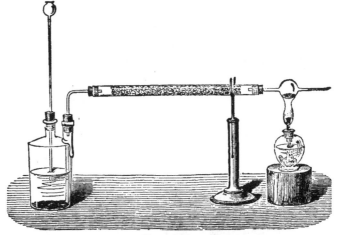

Fig. 123.

carbon separates as shining scales of graphite, and the iron is then softer, has a closer structure, and a gray fracture. It is called *gray iron*. Iron containing sulphur and phosphorus is always white; phosphorus renders iron brittle while cold, and sulphur renders it brittle while hot. In the first case the iron is said to be cold-short, while in the second it is called red-short.

Spiegeleisen, or looking-glass iron, is a variety of cast-iron very rich in carbon, and containing also manganese; it is employed in the manufacture of steel; it is crystalline, and breaks with smooth and highly-lustrous surfaces.

517. Steel is iron containing from 0.2 to 2 per cent. of carbon, and traces of nitrogen. It is obtained by a number of processes, which depend either on the partial decarbonization of cast-iron or on the introduction of the required proportion of carbon into soft

iron. When manganiferous cast-iron is maintained for a time melted under a layer of magnetic iron ore or ferruginous scoriæ, and the operation is arrested at the proper moment, the iron still retains a certain proportion of carbon, and *natural steel* is obtained.

Cement-steel, or *blister-steel,* is made by piling soft-iron bars between layers of charcoal in fire-clay boxes, which are then heated to redness in a furnace, and the temperature is maintained for several days ; the iron absorbs a certain proportion of carbon, and is converted into steel. As, however, the exterior of the bars will necessarily contain more carbon than the interior, the metal is rendered homogeneous by being melted in crucibles heated in a powerful wind-furnace. It then constitutes *cast-steel.*

The most important method of manufacture of steel is named, from its inventor, the Bessemer process. It consists in completely decarbonizing cast-iron and then adding sufficient cast-iron of the proper quality to give to the whole mixture the desired proportion of carbon. The operation is conducted in oval vessels of strong

iron plate lined with infusible fire-brick. This apparatus, which is called a converter, is supported on trunnions, so that it may swing back and forward on a horizontal axis. One of the trunnions is hollow, and communicates with a pipe passing partly around the converter and then leading to its bottom ; the fire-brick is here pierced with a number of holes, so that a blast of air may be forced up through the contents of the converter (Fig. 124).

FIG. 124.

The plant, as the whole of any manufactory is called, is established near blast-furnaces, and cupola furnaces, for melting the pig-iron, are constructed near the converter. Everything being ready, burning wood is thrown into the converter, which is then partly filled with coke, and the blast is turned on so that the fire-brick lining is heated to whiteness. The converter is then inverted; the coke is dumped out, and molten iron is run in from cupola furnaces. During the filling, the converter is kept in an inclined position, so that the tuyères for the passage of the blast may not become filled with the molten iron. The blast, which is under strong pressure, is now turned on, and the converter is rotated to an upright position. As the air bubbles up through the iron, the carbon, silicon, and other oxidizable elements present are consumed, and a brilliant flame rushes with a roaring noise from the mouth of the converter. When the carbon of the iron is burned out, the appearance of the flame undergoes a change which informs the workmen of the termination of the operation. The converter is then inclined, and the blast is arrested. In the mean time, the quantity of iron in the charge being accurately known, the exact quantity of spiegeleisen required to convert this charge into steel, has been melted in another cupola furnace; as soon as the blast is stopped, this molten spiegeleisen is run into the converter, which is agitated backward and forward in order that the contents may be perfectly mixed. The steel is then poured out into an enormous ladle, which is carried by a revolving crane over the circumference of a circle around which are arranged ingot-moulds, into which the steel is cast. The motions of the converter and of the crane, and the opening and closing of the blast, are effected from a distance by means of hydraulic machinery.

518. The valuable qualities of steel depend upon the ease with which it can be hardened or softened at pleasure, and the operations by which the change in hardness is brought about, constitute the processes of tempering. When heated and allowed to cool slowly, steel becomes soft and malleable like soft iron, but if it be heated to redness and then suddenly cooled by plunging it into cold water, it is rendered hard and brittle; it is, however, still elastic. When steel is to be hardened, it must either be rendered exceedingly hard and brittle by rapid cooling, or allowed to

o　　　　　　　　　27

become soft by slow cooling.   Intermediate degrees of hardness
are obtained by re-heating hard steel to temperatures depending on
the desired hardness.   Part of the temper is then said to be
drawn.   The color which the surface of the metal assumes is an
index of the temperature.

Straw yellow corresponds to 220°.
Brown   "   255°.
Light blue   "   285–290°.
Indigo blue   "   295°.
Sea green   ::   331°.

When a number of small objects are to be tempered alike, they
are heated in a bath of mercury or oil, of which the temperature
can be exactly regulated.

We may very well study the phenomena of tempering by heating a steel wire or a piece of watch-spring to redness and quickly
immersing it in water.   It is now very hard and brittle ; it breaks
as readily as a piece of glass.   We again heat it gently until its
surface becomes of a blue color, and now, whether we dip it in
water or allow it to cool slowly, we will find that it has become
quite elastic, but is still hard.   When, however, we heat it to redness and allow it to cool slowly, it becomes soft and flexible; it
will not break, but will retain any form into which it is bent.

The process of *casehardening*, which is applied to inferior
kinds of cutlery, consists in embedding the objects, which are
made of soft iron, in charcoal contained in crucibles or clay boxes.
These are then heated to bright redness, and the surface of the
iron becomes converted into steel.

Steel is less fusible than cast-iron, but much more fusible than
soft iron ; at the temperature at which soft iron becomes pasty,
steel melts.

LESSON LIX.

IRON AND ITS COMPOUNDS.

519. The density of soft iron varies from 7.4 to 7.9. It is ductile, malleable, and very tenacious. It fuses only at the highest temperatures of a powerful wind-furnace, but at a high white heat it becomes so soft that two pieces of the metal may be readily united in a solid mass by hammering or by pressure: the operation is called welding. At a somewhat lower temperature it may be readily rolled into sheets or bars, and sheet-iron is made by passing the heated metal between polished steel rollers. Iron may be rolled into leaves as thin as paper. Tin plate is sheet-iron covered with a coating of tin. Galvanized iron is made by dipping perfectly clean sheet-iron into melted zinc.

Iron is attracted by a magnet, and becomes itself a magnet while under the magnetic influence, but loses its magnetism when the exciting cause is removed. Under the same circumstances steel becomes a permanent magnet.

Unless in a state of fine division, iron is unaffected by dry air at temperatures below redness, but at a red heat it combines with oxygen and is converted into a black oxide which forms scales on its surface. It is rapidly rusted by moist air, and the rust is a hydrated ferric oxide.

When the formation of rust has begun, it proceeds with great rapidity, and if the mass of iron be of such a form that a large surface is combined with a comparatively small bulk, as in a long coil of wire or mass of small scrap iron partly immersed in water, the temperature may be much elevated by the rusting. It appears that hydrogen dioxide is formed during the rusting of iron, and that substance would greatly accelerate the change: the nitrogen of the air also plays some part in the phenomenon, for rust always contains a trace of ammonia.

At a red heat, iron decomposes water, liberating hydrogen, and forming an oxide. It is dissolved by hydrochloric and sulphuric acids, hydrogen being set free; this hydrogen has an unpleasant

odor, probably due to carbon compounds formed by the action
of the carbon of the iron.   Dilute nitric acid also dissolves iron,
disengaging red vapors, but the strongest nitric acid does not
affect it.

If some clean iron wire or some bright nails be dropped into pure nitric acid,
or a mixture of strong nitric and sulphuric acids, no action takes place; the
iron may now be removed and placed in more dilute acid, and even here it
will not dissolve: it is said to be in the passive state.   Its surface has become
covered with a protecting layer of gas derived from the strong acid: if while
the passive iron is immersed in the dilute acid we touch its surface with a
copper wire, the coating of gas is broken at one point, chemical action is at
once re-established, and the iron is quickly dissolved.

Iron forms two series of compounds,—ferric compounds, which
are analogous to those of aluminium, and in which two tetratomic
atoms pass from molecule to molecule as a hexatomic couple, and
ferrous compounds, in which it is more convenient to consider the
iron atom as diatomic.

520. FERROUS CHLORIDE, $FeCl^2$, is made by passing dry hydrochloric acid gas over metallic iron heated to redness in a porcelain
tube.   It then condenses in white, pearly scales in the cooler part
of the tube.

A solution of ferrous chloride may be obtained by dissolving iron in hydrochloric acid.   When the filtered liquid is sufficiently evaporated, it deposits
bluish-green crystals in which every molecule of ferrous chloride is combined
with four molecules of water.

521. FERRIC CHLORIDE, Fe^2Cl^6, sublimes in brilliant violet
crystals when chlorine is passed over incandescent iron contained
in a glass or porcelain tube.   It is very soluble in water, but its
solution undergoes a curious change by boiling.   A solution of
ferric chloride-is obtained by dissolving ferric oxide or powdered
hematite in hot hydrochloric acid.   When this solution is evaporated at a low temperature, the hydrated ferric chloride remains as
a brownish-yellow, deliquescent mass, but when the solution is
boiled its color darkens, and the reactions and general properties
of the liquid seem to show that it has been decomposed into
hydrochloric acid and a soluble variety of ferric hydrate.

522. FERROUS OXIDE, FeO, has been obtained as a black powder by passing

a mixture of carbon dioxide and carbon monoxide in equal volumes over heated ferric oxide. Carbon monoxide alone would yield metallic iron.

523. FERRIC OXIDE, Fe^2O^3, constitutes the minerals known as red hematite and specular iron. It is obtained as a fine red powder by strongly heating ferrous sulphate in a crucible: sulphur dioxide and sulphur trioxide are disengaged, while ferric oxide remains.

$$2FeSO^4 = SO^2 + SO^3 + Fe^2O^3$$

This powder is very hard, and is used for polishing under the names jewellers' rouge and *colcothar.*

When an alkaline hydrate or ammonia is added to a solution of ferric chloride, a flocculent, brown precipitate of ferric hydrate, $Fe^2(OH)^6$, is thrown down. This is the precipitate which, after being thoroughly washed, is the proper antidote for poisoning by arsenious oxide. Ferric solutions containing tartaric acid are not precipitated by the alkaline hydrates.

Rust is a ferric hydrate of which the composition usually corresponds with the formula $(Fe^2O^3)^23H^2O = Fe^2(OH)^6 + Fe^2O^3$. This is also the composition of the natural hydrate brown hematite. Goethite is a hydrate having the composition $Fe^2O^3.H^2O = Fe^2(OH)^2O$.

There is a soluble modification of ferric hydrate. It may be obtained by pouring a solution of ferric chloride which has been heated to 100° into the inner vessel of a dialyser (§ 220), the water in the exterior vessel being frequently changed. Hydrochloric acid passes through the membrane, while a solution of ferric hydrate remains within. Dialysis of a solution of ferric acetate yields soluble ferric hydrate in the same manner. This solution is used in medicine under the name dialysed iron.

524. FERROSO-FERRIC OXIDE, Fe^3O^4, is magnetic oxide of iron, commonly called black oxide of iron. It is found native in large quantities in the neighborhood of Lake Superior. It forms in black scales on the surface of iron heated to redness in the air. It is attracted by the magnet. It is a compound of ferrous and ferric oxides, $Fe^3O^4 = FeO.Fe^2O^3$.

525. FERROUS SULPHIDE, FeS, so largely used in the laboratory for the preparation of hydrogen sulphide, is made by heating a mixture of iron filings with two-thirds its weight of sulphur to redness in a covered crucible. After fusion, the mass is poured out, and on cooling forms a black solid of a metallic appearance.

526. IRON DISULPHIDE, FeS^2, constitutes the common mineral

iron pyrites. It is dimorphous, being found in cubical crystals of a yellow color and metallic lustre, known as yellow pyrites; and as rhombic prisms of a pale, greenish-yellow color, constituting white pyrites. When pyrites is heated in closed vessels, part of its sulphur distils; when it is heated in contact with air, the sulphur burns into sulphur dioxide, while the iron remains as oxide. The brilliant metallic appearance of iron pyrites has sometimes caused it to be mistaken for gold, and it has been called fool's gold: the action of heat at once reveals its true character .

527. FERRIC SULPHATE, $Fe^2(SO^4)^3$.—Ferrous sulphate, or green vitriol, has already been described (§ 123): when crystals of this salt are dissolved in water, and boiled with a little less than one-sixth their weight of sulphuric acid, and small quantities of nitric acid are added from time to time, a solution of ferric sulphate is obtained. When this liquid is evaporated to dryness, ferric sulphate remains as a yellowish-white mass, very soluble in water. By using a smaller quantity of sulphuric acid, various basic salts are obtained, and they may be considered as ferric sulphate in which one or two groups, SO^4, are replaced by as many atoms of oxygen. Such are $Fe^2O(SO^4)^2$ and $Fe^2O^2SO^4$. A mixture of these basic sulphates is employed in medicine under the name Monsel's solution. It is astringent and styptic, and is valuable for arresting hemorrhage.

528. TESTS FOR IRON.—The ferrous and the ferric salts are characterized by different reactions; by reducing agents such as nascent hydrogen produced by zinc and hydrochloric acid, the ferric salts are converted into ferrous salts, while ebullition with nitric acid, or the addition of chlorine-water, will produce a ferric compound from a ferrous salt.

Solutions of the ferrous salts are pale green; hydrogen sulphide occasions in them no precipitate, but ammonium sulphide throws down black ferrous sulphide. The alkaline hydrates and ammonia produce greenish-white precipitates of ferrous hydrate which rapidly become dark by absorbing oxygen from the air. Potassium ferrocyanide forms a pale-blue precipitate; potassium ferricyanide occasions a dark-blue precipitate, called Turnbull's blue.

Solutions of ferric salts are yellowish-brown or brown. With hydrogen sulphide they yield a precipitate of sulphur, being reduced to ferrous salts; ammonium sulphide throws down a black precipitate. The alkaline hydrates and ammonia form rust-colored precipitates of ferric hydrate, insoluble in an excess of the reagent. Potassium ferrocyanide throws down Prussian blue; potassium ferricyanide occasions no precipitate. Potassium sulphocyanate produces a blood-red color, due to the formation of ferric sulphocyanate. Tannin, or an infusion of gall-nuts, forms a blue-black and very finely divided precipitate, which long remains suspended in the liquid.

LESSON LX.

COBALT, NICKEL, AND MANGANESE.

529. Cobalt, $Co = 59$.—This metal is found combined with sulphur and arsenic; the mineral *cobaltine* is a sulpharsenide of cobalt, having the composition $CoSAs$. Metallic cobalt is obtained by strongly heating with a little charcoal in a covered crucible the cobalt oxalate, which is precipitated by the addition of ammonium oxalate to the solution of a cobalt salt. Carbon dioxide is disengaged, while cobalt remains as a dull powder, which may be fused into a button by the highest heat of a wind-furnace. It is a silvery-white metal, and is malleable and ductile. It is attracted by the magnet. Its density is 8.6. It is unaffected by either dry or moist air at ordinary temperatures, but is oxidized at a red heat.

Cobalt forms cobaltous oxide, CoO, a sesquioxide, Co^2O^3, and several other oxides which appear to be formed by a combination of these two in different proportions.

530. COBALT CHLORIDE, $CoCl^2$, is prepared by dissolving either the oxide or carbonate of cobalt in hydrochloric acid. The solution is red, and, when concentrated, deposits red crystals containing six molecules of water of crystallization. Anhydrous cobalt chloride is blue: if a little strong sulphuric or hydrochloric acid be added to a concentrated solution of cobalt chloride, the

liquid becomes blue. It contains anhydrous cobalt chloride. Writing made on paper with a very dilute solution of cobalt chloride is invisible when dry ; the small quantity of the salt present is hydrated ; but if the paper be heated, the characters become blue, for the water is driven off. After exposure to the air for a time, the characters again fade, the cobalt chloride absorbing atmospheric moisture.

531. COBALT BLUE —The ores of cobalt are principally employed for the manufacture of a dark-blue substance generally called *smalt*. This is a mixture of cobalt silicate and potassium silicate. It is prepared by partially roasting the ore in order to convert the greater part of the cobalt into oxide. The roasted mass is then pulverized and melted with a mixture of potassium carbonate and white quartz sand. A blue, vitreous mass is thus obtained, which floats on a fused mass containing the iron, nickel, copper, and unaltered sulphur and arsenic of the ore. This mixture has a metallic appearance ; it is called *speiss*, and is used for the preparation of nickel. While still molten, the blue glass constituting smalt is poured into water, in which it breaks up into small fragments, which are readily pulverized.

An impure sesquioxide of cobalt is used for painting on glass and porcelain, to which it communicates a deep-blue color.

532. TESTS FOR COBALT.—The more ordinary salts of cobalt form rose-colored or currant-red solutions, but if these solutions contain free acid, they become blue when heated. They are not precipitated by hydrogen sulphide; ammonium sulphide forms a black precipitate. The alkaline hydrates produce blue precipitates, which an excess of the reagent converts into rose-colored cobaltous hydrate, $Co(OH)^2$. Ammonia-water occasions a blue precipitate, which dissolves in an excess of the reagent, an ammonio-cobalt salt being formed. When strongly heated with a little borax on the end of a platinum wire, the compounds of cobalt yield beads of a blue glass.

533. **Nickel,** $Ni = 59$.—Nickel is found principally as arsenide, $NiAs^2$, in the mineral *kupfernickel*, which is generally associated with ores of copper and iron. It is obtained from this mineral and from the speiss formed during the manufacture of smalt. The speiss or kupfernickel is first roasted, in order that the nickel may be converted into oxide, and is afterwards heated with powdered charcoal as long as arsenic vapors are disengaged. The mass is then dissolved in nitro-hydrochloric acid, and the solution is evaporated to expel the excess of acid. Hydrogen sul-

phide is passed through the liquid as long as it occasions any precipitate, and all the metals present excepting iron, nickel, and cobalt are thus precipitated as sulphides. The addition of sodium carbonate to the clear decanted liquid precipitates the iron, nickel, and cobalt. The mixed precipitate is washed and treated with oxalic acid and an excess of ammonia-water. An ammoniacal solution of the oxalates of nickel and cobalt is so obtained; this is exposed to the air, and as the ammonia gradually volatilizes, the nickel oxalate is first deposited in greenish crystals, after which the cobalt oxalate separates. The precipitates are removed from day to day, and the nickel salt is several times dissolved in ammonia, and allowed to separate until it is free from cobalt. It is then treated with an alkaline hydrate, and the resulting nickel hydrate is dried, mixed with charcoal powder and oil, and the paste thus obtained is calcined at the highest temperature of a good furnace. The reduced nickel then fuses and forms a mass.

Nickel may be prepared in the laboratory by strongly heating its oxalate out of contact with the air.

Nickel is a yellowish-white metal, capable of taking a high polish. It is malleable, ductile, and very tenacious. Its density is about 8.5. It is attracted by the magnet. It is the hardest of the more common metals. It is not affected by the air at ordinary temperatures, but becomes oxidized at a red heat. It is slowly dissolved by dilute hydrochloric and sulphuric acids, more rapidly by nitric acid.

Nickel is employed in the manufacture of a number of alloys. German silver contains 25 per cent. of nickel, 25 per cent. of zinc, and 50 per cent. of copper, but the proportions vary greatly. The white nickel coins of the United States contain 25 per cent. of nickel and 75 per cent. of copper.

Nickel is largely employed for plating articles of brass, iron, and steel, and its hardness, its high lustre, and its freedom from rust render it admirably adapted to this purpose. The well-cleaned objects are attached to the zinc pole of a voltaic battery and immersed in a solution of nickel and ammonium double sulphate: the positive pole of the battery is connected with a plate of pure nickel dipped in the same liquid.

v

534. NICKEL CHLORIDE, $NiCl^2$, is made by dissolving the oxide or hydrate in hydrochloric acid. When sufficiently concentrated, the green solution deposits green crystals containing $NiCl^2 + 6H^2O$.

535. NICKEL MONOXIDE, NiO, is a pale-gray powder, obtained by strongly heating the carbonate or nitrate. Nickel hydrate, $Ni(OH)^2$, is thrown down as a pale-green precipitate when an alkaline hydrate is added to the solution of a nickel salt. When chlorine is passed through water in which this precipitate is suspended, a hydrate of nickel sesquioxide, Ni^2O^3, is formed.

536. NICKEL SULPHATE, $NiSO^4$.—When nickel oxide or hydrate is dissolved in dilute sulphuric acid, and the solution is allowed to evaporate spontaneously, green crystals of the sulphate with seven molecules of water of crystallization are deposited. With ammonium sulphate, this compound forms a double salt in fine bluish-green crystals containing $NiSO^4.(NH^4)^2SO^4 + 6H^2O$. This is the salt used in nickel-plating.

537. TESTS FOR NICKEL.—The anhydrous nickel salts are yellow, but the crystallized salts and their solutions are emerald-green. If the solution be acid, hydrogen sulphide produces no precipitate, but a black precipitate of sulphide is thrown down if the solution contain sodium acetate. The same precipitate is formed by ammonium sulphide. The alkaline hydrates and carbonates occasion pale-green precipitates. Ammonia-water forms a green precipitate, which quickly dissolves in an excess of the reagent, yielding a blue solution.

538. MANGANESE, Mn $= 55$.—This is an exceedingly infusible metal, so hard that it will scratch steel. It has been obtained by strongly heating a mixture of manganoso-manganic oxide and sugar. Three of its oxides are found native in the minerals *braunite*, Mn^2O^3, *pyrolusite*, MnO^2, and *hausmannite*, Mn^3O^4.

539. MANGANESE DIOXIDE, MnO^2.—This compound is commonly called black oxide of manganese. When it is heated to redness, it loses one-third of its oxygen, and is converted into a red powder of manganeso-manganic oxide, Mn^3O^4. This decomposition was formerly applied for the preparation of oxygen.

$$3MnO^2 = Mn^3O^4 + O^2$$

Oxygen is also evolved, while manganous sulphate is formed, when the dioxide is heated with sulphuric acid.

$$2MnO^2 + 2H^2SO^4 = 2MnSO^4 + 2H^2O + O^2$$

When heated with hydrochloric acid, manganese dioxide yields manganese chloride, water, and chlorine: large quantities of the dioxide are used for the manufacture of chlorine by this reaction, and in the solution of manganese chloride the dioxide is regenerated by various processes. One of these consists in mixing the liquid with milk of lime, by which calcium chloride and manganous hydrate, $Mn(OH)^2$, are formed: heated air is then blown through the mixture, and the manganous hydrate is converted into the dioxide, from which the solution of calcium chloride is decanted.

Manganese dioxide is also employed to decolorize glass rendered dark by carbonaceous matter or green by a trace of iron. In the first case it oxidizes the carbon in the ‍glass-pot, and in the second it neutralizes the green color of the ferrous silicate by imparting a reddish tint of manganous silicate.

540. MANGANIC ACID, H^2MnO^4.—When strongly heated with alkaline hydrates, manganese dioxide absorbs oxygen from the air, and an alkaline manganate is formed. A mixture of manganese dioxide and potassium hydrate may be fused in a silver or iron dish, and, when the cold mass is treated with water, a green solution is obtained. If this be evaporated at a low temperature in a vacuum, it deposits green crystals of potassium manganate.

When the alkaline manganates are heated to 450° in a current of steam, they are decomposed into alkaline hydrate and manganese dioxide. This decomposition has been applied to the manufacture of oxygen on a large scale. A mixture of sodium hydrate and manganese dioxide is heated in a current of air; oxygen is absorbed, and sodium manganate is formed.

$$MnO^2 \quad + \quad 2NaOH \quad + \quad O \quad = \quad Na^2MnO^4 \quad + \quad H^2O$$
Manganese dioxide.   Sodium manganate.

The air is then stopped, and steam is passed over the heated manganate, reproducing sodium hydrate and manganese dioxide, while oxygen is disengaged.

$$Na^2MnO^4 \quad + \quad H^2O \quad = \quad 2NaOH \quad + \quad MnO^2 \quad + \quad O$$

The oxygen, with the excess of steam, is led through cold pipes, where the steam is condensed, while the oxygen passes on to appropriate gas-holders.

541. PERMANGANATES.—When the green solution of potassium manganate is boiled, its color changes to red, while hydrated manganese dioxide separates in brown flakes. The red color is due

to the formation of potassium permanganate, and the solution contains free sodium hydrate.

$$3K^2MnO^4 \quad + \quad 2H^2O \quad = \quad\quad K^2Mn^2O^8 \quad + \quad MnO^2 \quad + \quad 4KOH$$
Potassium manganate.   Potassium permanganate.

A similar reaction takes place when an acid is added to the solution of a manganate.

Potassium permanganate is made by heating in an iron crucible a mixture of five parts of potassium hydrate with a little water, and three and a half parts of potassium chlorate with four of manganese dioxide in fine powder. The temperature is gradually raised to dull redness, the mass being constantly stirred. It is allowed to cool, and, after being pulverized, is thrown into two hundred parts of boiling water, and stirred until the liquid has assumed a purple color. It is then left to settle; the clear liquid is decanted, neutralized with nitric acid, and evaporated on a water-bath. The crystals which separate on cooling are drained on a clean brick.

They are purple-black needles, having a metallic reflection, soluble in about fifteen times their weight of cold water. The solution has an intense purple color. Potassium permanganate is an energetic oxidizing agent; its solution is at once decolorized by sulphur dioxide, which it converts into sulphuric acid, and the liquid contains sulphuric acid, potassium sulphate, and manganous sulphate.

$$K^2Mn^2O^8 \quad + \quad 5SO^2 \quad + \quad 2H^2O \quad = \quad K^2SO^4 \quad + 2MnSO^4 \quad + \quad 2H^2SO^4$$
Potassium permanganate.   Manganous sulphate.

The oxidizing properties of potassium permanganate are largely employed in the laboratory.

542. TESTS FOR MANGANESE.—The salts of manganese are colorless or pale rose-colored. They are not precipitated by hydrogen sulphide, but ammonium sulphide throws down flesh-colored manganese sulphide. The alkaline hydrates produce dirty-white precipitates of manganese hydrate, which soon absorbs oxygen from the air and becomes brown. When heated with a little potassium hydrate or nitrate or sodium carbonate on a piece of platinum foil, they yield a greenish mass of an alkaline man-

ganate, which forms a red solution when treated with a little
dilute nitric acid.

Uranium is an element closely related to manganese. One of its principal
minerals is *pitchblende*, a greenish-yellow oxide. It is used for the preparation of sodium uranate, which is known as uranium yellow; this is employed
for painting on porcelain, and for coloring a greenish-yellow glass which is
highly fluorescent.

LESSON LXI.

CHROMIUM AND TIN.

543. Chromium, Cr = 52.5.—Chromium exists in the mineral
chromite, which is a compound of chromium oxide and ferrous
oxide, and may be considered as ferroso-ferric oxide in which the
ferric oxide is replaced by the sesquioxide of chromium, $FeO.Cr^2O^3$.
It is found also as lead chromate in the red lead of Siberia. The
metal has been obtained by calcining a mixture of its oxide with
charcoal and oil. It then forms grayish-white, metallic granules,
which are exceedingly hard.

544. CHROMIC CHLORIDE, Cr^2Cl^6, is obtained by passing chlorine gas over
an incandescent mixture of chromium sesquioxide and charcoal; it then sublimes and condenses in brilliant violet scales in the cooler parts of the tube.
By the action of hydrogen at a red heat, it is converted into white chromous
chloride, $CrCl^2$. Chromic chloride is insoluble in water, but dissolves readily
in presence of a small quantity of chromous chloride, yielding a green solution
from which there may be obtained a crystallized hydrate, $Cr^2Cl^6 + 6H^2O$.

545. CHROMIUM SESQUIOXIDE, Cr^2O^3.—When potassium dichromate is heated
in a crucible with about half its weight of sulphur, a mass is obtained from
which water dissolves potassium sulphate, leaving chromium sesquioxide as a
green powder. Instead of sulphur, starch in quantity equal to one-fourth
the weight of the dichromate may be employed, but the resulting oxide must
afterwards be recalcined in the air, to burn out traces of carbon. It is in the
latter manner that the fine chrome green used for painting on porcelain is obtained. Chromium sesquioxide is not decomposed by heat, and fuses only at
very elevated temperatures. A corresponding hydrate, $Cr^2(OH)^6$, is thrown
down as a bluish-green precipitate when ammonia-water is added to the
green solution of chromic chloride. The same hydrate is precipitated by the
alkaline hydrates, but dissolves in an excess of the reagent; when the liquid
is boiled, the insoluble oxide is thrown down.

546. CHROMIC ANHYDRIDE, CrO^3.—This compound is commonly called chromic acid. It is prepared by mixing a cold saturated solution of potassium dichromate with one and a half times its volume of strong sulphuric acid. As the liquid cools, chromium anhydride separates in crimson needles, which are quickly drained on a dry brick and recrystallized in the smallest possible quantity of warm water. It is a deliquescent substance, exceedingly soluble in water, and the solution has an orange color. It energetically oxidizes many bodies. With hydrochloric acid it forms water and chromium sesquichloride, while chlorine is set free.

$$2CrO^3 + 12HCl = Cr^2Cl^6 + 6H^2O + 3Cl^2$$

It instantly oxidizes sulphur dioxide, chromium sulphate being formed.

$$2CrO^3 + 3SO^2 = Cr^2(SO^4)^3$$

It oxidizes alcohol and ether with such energy that those compounds are inflamed.

547. CHROMATES.—The solution of chromium anhydride must be regarded as containing chromic acid, $H^2CrO^4 = H^2O + CrO^3$, corresponding in molecular constitution to sulphuric acid. The chromium compounds are all derived from potassium dichromate, which is manufactured from chrome iron. The pulverized chromite is heated to redness with half its weight of potassium nitrate, and the mass is then exhausted with water. An impure solution of potassium neutral chromate is thus obtained: this is treated with acetic acid, which combines with part of the potassium, converting the neutral chromate into dichromate; the clear solution of potassium dichromate is then decanted and evaporated until it is ready to crystallize.

548. *Potassium chromate*, K^2CrO^4, forms beautiful, lemon-yellow, anhydrous crystals, which are very soluble in water, to which they impart an intense yellow color.

549. *Potassium dichromate*, $K^2Cr^2O^7$, forms large, orange-red crystals, soluble in about eight times their weight of cold water, and in much less boiling water. By heat they are decomposed into potassium chromate, chromium oxide, and oxygen.

$$2K^2Cr^2O^7 = 2K^2CrO^4 + Cr^2O^3 + O^3$$

Potassium dichromate is an energetic oxidizing agent. When heated with sulphuric acid, it yields oxygen, while chrome alum will crystallize from the liquid obtained by treating the residue with boiling water.

$$K^2Cr^2O^7 + 4H^2SO^4 = Cr^2(SO^4)^3.K^2SO^4 + 4H^2O + O^3$$

When sulphur dioxide is passed into a solution of potassium dichromate, the orange color is gradually replaced by green : while the sulphur dioxide becomes sulphuric acid, both chromium and potassium are converted into sulphates, and the liquid will yield chrome alum if sulphuric acid be added.

$$K^2Cr^2O^7 + 3SO^2 + H^2SO^4 = Cr^2(SO^4)^3.K^2SO^4 + H^2O$$

550. *Ammonium dichromate*, $(NH^4)^2Cr^2O^7$, may be made by dividing a solution of chromic acid into two equal portions, neutralizing one with ammonia-water, and then adding the other. When the solution is evaporated, the ammonium dichromate separates in red crystals, which, when heated, yield pure chromium trioxide in a curious pulverulent form, resembling green tea.

$$(NH^4)^2Cr^2O^7 = Cr^2O^3 + 4H^2O + N^2$$

551. *Lead chromate*, $PbCrO^4$, is found native as the red lead of Siberia. It is made by mixing solutions of potassium chromate or dichromate and lead acetate; potassium acetate remains in solution, while the lead chromate forms a dense yellow precipitate, which when washed and dried constitutes chrome yellow. It is insoluble in water and in acetic acid, but dissolves in solutions of the alkaline hydrates. It melts at a red heat, and is readily reduced by both hydrogen and charcoal. It is sometimes substituted for cupric oxide in the analysis of carbon compounds.

552. TESTS FOR CHROMIUM.—Although there is a series of chromous salts in which the chromium atom is diatomic, the reactions of the salts corresponding to the sesquioxide are sufficient to characterize this element. In the green solutions of these compounds hydrogen sulphide produces no precipitate : ammonium sulphide throws down chromium hydrate, hydrogen sulphide being disengaged.

$$Cr^2Cl^6 + 3(NH^4)^2S + 6H^2O = Cr^2(OH)^6 + 6NH^4Cl + 3H^2S$$

The alkaline hydrates and ammonia produce the same green precipitate, which dissolves readily in an excess of the former reagents, more slowly in ammonia-water. When the solution so obtained is boiled, anhydrous chromium sesquioxide is precipitated, and does not redissolve on cooling.

The chromates may be identified by heating them in a test-tube with a little common salt and sulphuric acid. Irritating red vapors of an oxychloride of chromium, CrO^2Cl^2, are disengaged, and if conducted into a cold tube will condense to a blood-red liquid. When passed into water, these vapors are decomposed into hydrochloric and chromic acids, and the liquid will yield yellow lead chromate when treated with a solution of lead acetate.

$$CrO^2Cl^2 \; + \; H^2O \; = \; CrO^3 \; + \; 2HCl$$

553. Closely related to chromium in their chemical relations are the elements molybdenum and tungsten, the first of which exists as sulphide, MoS^2, in the mineral *molybdenite,* while the second is found in various tungstates, as in the mineral *wolfram,* which is a tungstate of iron and manganese. Both of these elements form trioxides, which, like chromium trioxide, are the anhydrides of corresponding acids.

554. **Tin,** $Sn = 118$.—Tin is rarely found in the metallic state in nature: its only workable ore is the dioxide, which constitutes the mineral *cassiterite.* The principal tin-mines are in Wales. The ore is crushed, and the dioxide, being very heavy, can be separated from the lighter earthy matters by washing in a stream

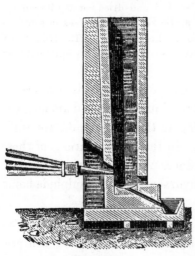

of water. It is then roasted, the sulphides and arsenides of iron and copper present being converted into oxides, which are removed by a second washing. The purified cassiterite is mixed with charcoal and fed into a cupola furnace (Fig. 125), where the combustion is supported by a blast of air. The reduced tin collects on the hearth of the furnace, and runs into a basin, where it is stirred with poles of green wood. The gases given off reduce any oxide that has been formed, and bring to the surface of the molten metal

FIG. 125.

the foreign matters, which form a dross. The tin is then drawn off into moulds, and is purified by being melted at a low temperature on the inclined hearth of a reverberatory furnace. Being more fusible than the foreign metals present, it melts first, and runs into a cavity prepared for it, while the less fusible metals remain on the hearth.

Tin is a silvery-white metal, having a density of about 7.3. It melts at 228°, and may be crystallized by slow cooling. It is malleable and ductile : when a bar of tin is bent, it produces a peculiar noise, called the cry of tin, caused by the sliding of the crystals over one another.

It is not affected by the air at ordinary temperatures, but when melted absorbs oxygen, and by stirring may be entirely converted into the dioxide. It is dissolved by hydrochloric acid, hydrogen being disengaged, while stannous chloride is formed. Nitric acid converts tin into dioxide, giving off torrents of red vapors. Hot solutions of the alkaline hydrates dissolve tin, forming alkaline stannates, and disengaging hydrogen.

Tin is used for the manufacture of tin foil, employed for enveloping tobacco, chocolate, etc. ; also for tinning copper and iron, which is accomplished by dipping the perfectly clean objects into a bath of molten tin. Its resistance to the action of vegetable acids renders it invaluable as a coating for culinary utensils. It enters into the composition of plumbers' solder, which is an alloy of tin and lead. Bronze, bell-metal, gun-metal, and speculum-metal are alloys of tin and copper (page 287). Britannia metal is tin alloyed with a small proportion of antimony, bismuth, and copper.

In its compounds tin is either diatomic or tetratomic. Those in which it is diatomic are called stannous compounds, while in the stannic compounds it is tetratomic, one atom of tin having the same combining power as four atoms of hydrogen.

555. STANNOUS CHLORIDE, $SnCl^2$.—Anhydrous stannous chloride is obtained by passing hydrochloric acid gas over heated tin. It is a white solid, fusible at 250°. When it is dissolved in a small quantity of water, or when metallic tin is dissolved in hot hydrochloric acid, a solution is obtained which when sufficiently concentrated deposits crystals of a hydrate containing $SnCl^2 +$

$2H^2O$. These crystals are known in commerce as tin salt or tin crystals. They are soluble in a small quantity of water, but when the solution is diluted, a deposit of an oxychloride is formed containing $SnCl^2.SnO$. At the same time a certain proportion of the stannous chloride is converted into stannic chloride, $SnCl^4$. This decomposition is prevented by the presence of free hydrochloric acid, or by a small quantity of ammonium chloride. Stannous chloride is a reducing agent: a few drops of its solution instantly decolorize the purple solution of potassium permanganate; it reduces the salts of silver and gold, setting free the metal. When it is added to a solution of mercuric chloride, a white precipitate of mercurous chloride is formed, which an excess of stannous chloride converts into a gray deposit of finely-divided metallic mercury. In these reactions the stannous chloride becomes stannic chloride. Stannous chloride is used as a mordant in dyeing.

556. STANNIC CHLORIDE, $SnCl^4$, is formed with the production of light and heat by the direct union of tin and chlorine. It is prepared by passing dry chlorine over melted tin contained in a retort; it then distils, and condenses as a heavy, fuming, yellow liquid, which boils at 120°. It combines energetically with water, forming crystals of a hydrate containing $SnCl^4 + 5H^2O$. The same hydrate may be made by dissolving tin in hydrochloric acid and from time to time adding small quantities of nitric acid. The crystals are soluble in water, yielding a limpid solution.

557. STANNIC OXIDE, SnO^2.—When an alkaline hydrate is added to a solution of stannous chloride, stannous hydrate is formed as a white precipitate, which, by boiling, is converted into black stannous oxide, SnO. The addition of ammonia to a solution of stannic chloride throws down a white gelatinous precipitate of stannic hydrate, H^2SnO^3, which by the action of heat is converted into stannic oxide, SnO^2. This compound is found in nature in hard, transparent crystals; it is cassiterite. It is an acid oxide, and stannic hydrate reacts with the bases, forming stannates whose compositions correspond to the carbonates. The white powder produced by the action of nitric acid on tin is a stannic hydrate, containing $Sn(OH)^4 = SnO^2 + 2H^2O$.

558. *Sulphides of Tin.*—By heating together the proper proportions of tin and sulphur, two sulphides may be obtained. Stannous sulphide, SnS, is a gray, crystalline mass. The preparation of stannic sulphide, SnS^2, requires particular precautions; an amalgam of tin with half its weight of mercury is mixed with flowers of sulphur and ammonium chloride and heated to dull redness. Mercuric sulphide, ammonium chloride, and the excess of sulphur sublime, while the interior of the vessel becomes lined with a golden-yellow,

crystalline mass of stannic sulphide. The operation must then be arrested, or this compound will be decomposed into stannous sulphide and sulphur; the addition of the mercury and ammonium chloride is intended to keep the temperature down to the volatilizing points of mercuric sulphide and ammonium chloride. Stannic sulphide forms soft, crystalline scales, called mosaic gold.

559. TESTS FOR TIN.—In stannous solutions, both hydrogen sulphide and ammonium sulphide form brown precipitates, soluble in an excess of ammonium sulphide. The alkaline hydrates and ammonia give white precipitates, soluble in an excess of the former reagents, but insoluble in ammonia. Gold trichloride throws down purple of Cassius. In mercuric chloride solutions, an excess of stannous chloride precipitates gray metallic mercury.

In stannic solutions, hydrogen sulphide and ammonium sulphide form yellow precipitates, soluble in a large quantity of the latter reagent. A piece of zinc placed in either a stannous or a stannic solution becomes covered with a deposit of tin, which may be rendered brilliant by burnishing.

560. The elements titanium, zirconium, and thorium closely resemble tin in their chemical relations, although their physical properties are very different. Each forms a tetrachloride and a dioxide. Titanium is found native as dioxide in the minerals *rutile, anatase,* and *brookite.* Zirconium occurs as a silicate in the mineral *zircon.*

LESSON LXII.

PLATINUM AND ITS ALLIED METALS.

561. Like gold, platinum is found in the metallic state in rounded granules distributed through sandy deposits. Being very heavy, it is also separated like gold, by washing the sand in a stream of water. The native platinum, however, is not pure: besides containing traces of gold, copper, and iron, it is alloyed with several other metals which it resembles in certain properties, and which are called the platinum metals. They are rhodium, ruthenium, palladium, iridium, and osmium. The platinum is extracted by treating the grains first with dilute nitro-hydrochloric

acid, which removes all excepting the platinum metals, and then heating it with strong nitro-hydrochloric acid, which dissolves the platinum, leaving osmium and the greater part of the iridium. The liquid is then exactly neutralized with sodium carbonate, and a solution of mercuric cyanide is added. This throws down a precipitate of palladium cyanide, which is removed by filtration, and the clear liquid is treated with ammonium chloride. A crystalline precipitate of a double chloride of platinum and ammonium forms, and this, when calcined, leaves a porous gray residue of platinum sponge. Platinum so prepared always contains some iridium, for the latter metal also separates as a double chloride when the ammonium chloride is added.

In order to agglomerate the spongy platinum, it is made into a stiff paste with a little water, and this is strongly compressed in a slightly conical steel cylinder. It is then removed, heated to whiteness, and converted into a solid mass by hammering. Platinum is also melted in lime crucibles, heated by the flame of the oxyhydrogen blow-pipe directed against the mass of metal.

Platinum is a white metal, having a somewhat gray lustre. Its density is 21.5. It is very malleable and ductile. It melts only at the highest attainable temperatures, but at a white heat it becomes soft and can be forged and welded like iron. It is not affected by the air at any temperature, and does not dissolve in either hydrochloric, sulphuric, or nitric acid. When it is alloyed with silver, it is dissolved by nitric acid. Nitro-hydrochloric acid dissolves it slowly in the cold, more rapidly by the aid of heat, converting it into the tetrachloride. The alkaline hydrates and nitrates attack it at a high temperature, and these substances must not be fused in platinum crucibles.

Platinum has the power of condensing gases in its pores, and we have already seen how the oxidation of ammonia and vapor of alcohol and ether may be effected by a platinum wire. This property is more strongly manifested by platinum sponge than by the compact metal, and hydrogen escaping from a jet may be ignited by holding in it a morsel of recently-heated spongy platinum. When a solution of platinic chloride is boiled with potassium hydrate, and alcohol is added to the boiling liquid with constant stirring, metallic platinum is deposited as a black powder. This powder, which is called *platinum black*, is in a state of extreme division, and brings about the oxidation of combustible gases and vapors even more readily than platinum sponge.

Platinum is employed for the manufacture of crucibles and dishes for the laboratory, for it not only resists high temperatures but is attacked by very few chemical reagents. It is manufactured into large retorts for the concentration of sulphuric acid. All of this apparatus is made as thin as is consistent with strength, for the metal is costly.

Platinum forms two series of compounds,—platinous compounds, in which it is diatomic, and platinic compounds, in which it is tetratomic.

562. PLATINIC CHLORIDE, $PtCl^4$, is made by dissolving the metal in nitro-hydrochloric acid. When the reddish-brown liquid is sufficiently concentrated, it deposits, on cooling, hydrated crystals of platinic chloride, which may be rendered anhydrous by heat. The anhydrous salt is a red-brown deliquescent mass, very soluble in water, alcohol, and ether. When its solution is added to solution of potassium chloride or ammonium chloride, a yellow crystalline precipitate of a double chloride of platinum with potassium or ammonium is formed. These salts are not very soluble in cold water, and are still less soluble in alcohol, the potassium compound being more soluble than that of ammonium. Their compositions are $PtCl^4.2KCl$ and $PtCl^4.2NH^4Cl$. When the ammonium salt is heated, it leaves a residue of spongy platinum.

If platinum tetrachloride be carefully heated to 200°, chlorine is disengaged; and if the residue be extracted with boiling water, the unaltered platinic chloride is dissolved, while platinum dichloride, $PtCl^2$, remains as an olive-green powder.

Of the other metals of the platinum group, OSMIUM is the least fusible: it has never been melted. When strongly heated in the air, it forms a volatile oxide, which is one of the most dangerous poisons known. It is the heaviest known element, having a density of 22.48.

563. PALLADIUM is the most fusible of these metals, and has the lowest density, its specific weight being 11.4. When a piece of this metal is made

the negative electrode of an apparatus in which water is being decomposed by the voltaic current, it will absorb about nine hundred times its volume of hydrogen.

564. IRIDIUM often constitutes a considerable proportion of platinum ore, which is then called platiniridium or osmiridium, as platinum or osmium preponderates in the alloy. Iridium alloyed with 90 per cent. of platinum is as hard and elastic as steel, is less fusible than platinum, and is unaltered by the air. It is used for the points of gold pens and ink-containing pencils, as is also the native alloy, osmiridium. The density of iridium is 22.38.

565. RHODIUM is more fusible than iridium, but less fusible than platinum. Its density is 12.1. It does not dissolve in nitro-hydrochloric acid unless it is alloyed with other metals.

566. RUTHENIUM is the most infusible metal after osmium. Its density is 12.26. It is hardly attacked by boiling nitro-hydrochloric acid.

LESSON LXIII.

THE CHEMISTRY OF LIFE.

567. Under the influence of the mysterious principle which we call life, certain chemical compounds undergo a complete metamorphosis. Their elements become rearranged in manners which are beyond our methods of research, and the matter becomes *organized*. It assumes certain definite forms which we call cells, and living cells are gifted with a wonderful power of reproduction : under the proper conditions they can convert unorganized dead matter into other cells, either of the same kind or of very different kinds related by progressive modifications. Chemists are able to change one form of matter into another,—to modify and destroy molecules, and to construct new molecules; they are unable to create the simplest cell, the lowest form of organized matter. We have no reason to believe that any cell is ever produced except from another, but we know that under modified circumstances the nature of the cell may in the course of successive reproductions become completely modified, and new forms of organized matter are produced.

The elements which enter into the composition of organized

matter are comparatively few: all vital tissues contain carbon, hydrogen, and oxygen, and these three, together with nitrogen and a few salts, principally phosphates, chlorides, and sulphates of sodium, potassium, and calcium, constitute the greater part of all tissues, vegetable and animal. Plants and vegetables directly convert carbon dioxide and water into cellulose, starch, hydrates of carbon, and other compounds, among which are even hydrocarbons. In this reducing action of vegetable life on carbon dioxide and water, the atoms of carbon and hydrogen recover part of the energy which disappears from them in the formation of those compounds. As far as their matter is concerned, plants then act as storers or regenerators of energy. In the natural heat and motion of animals the atomic energy of the compounds of carbon and hydrogen is manifested as those compounds are again oxidized with the formation of carbon dioxide and water. Animal life is really dependent on the continual expenditure of energy. Vegetables can receive their nutrition directly from mineral matter, but animals can form tissues only from matter that has first been prepared by vegetables or other animals.

However, besides the carbon, hydrogen, and oxygen, plants absorb nitrogen from nitrogenized matters in the soil, and the nitrogen compounds of plants are essential for the nutrition of animals. We must study some of the compounds which have thus far been formed only under the influence of life, and of which the organization results in the production of cells. Among these substances, which we must remember are not chemical compounds of definite and known constitution, of first importance are the albuminoid matters.

ALBUMINOID AND GELATINOID SUBSTANCES.

568. These complex matters are composed of carbon, hydrogen, oxygen, and nitrogen, and often a small proportion of sulphur. By their compositions and properties, they are all related to the albumen of white of egg, or to the gelatin or glue which can be extracted from bones. If flour made from wheat or other cereal be kneaded in water, the starch is washed out, while a gray elastic

mass of gluten remains. This gluten may be separated into several different substances, having different degrees of solubility in alcohol: they are of similar composition, containing a little more than 50 per cent. of carbon, 7 of hydrogen, 17 of nitrogen, 20 of oxygen, and rather less than one per cent. of sulphur. The water used in the preparation of gluten contains another matter, which may be separated by allowing the starch to settle, adding a few drops of acid to the clear liquid, and heating to the boiling point. An albuminoid matter then coagulates in white flakes. Its composition does not differ greatly from that of the bodies separated from gluten, but there are slight differences depending on the grain or seed from which the substance is derived. From the seeds of leguminous vegetables, such as peas, beans, and lentils, a body called *legumine* may be extracted, and, in addition to the elements contained in gluten and other vegetable albumens, this substance contains a small percentage of phosphoric acid, probably in the form of a substituted acid, in which various carbon groups replace one or more hydroxyl groups of orthophosphoric acid.

The albuminoid matters of animals, which are derived from the similar vegetable substances, are classified more with reference to their behavior under the action of heat than according to their composition, which varies but little. They may, however, be arranged in two groups,—albuminoid matters and gelatin-like compounds.

The general composition of these bodies is as follows :

	Albumen Group.	Gelatin Group.
Carbon	53.5	50.0
Hydrogen	6.9	6.6
Oxygen	23.0	26.1 to 23.1
Nitrogen	15.6	16.8
Sulphur	1.0	0.5 to 3.5

Of the albuminoid matters we can consider only albumen, fibrin, casein, and hemoglobin.

569. ALBUMEN exists in a soluble form and an insoluble modification. Soluble, it occurs in white of egg, and in the serum or clear liquid of blood; but even these forms present certain differences. If either of these liquids be evaporated at a low temper-

ature, the albumen remains as a transparent, yellowish, gum-like mass, which is perfectly soluble in water. It is not pure, but contains a small quantity of alkaline carbonate and certain salts. If a solution of albumen be heated to 70°, it becomes clouded, and at a few degrees higher the albumen separates either in flakes or in a solid mass, according to the concentration of the solution. The soluble albumen has coagulated and has become insoluble albumen. Solutions of albumen are also coagulated by the addition of either sulphuric, nitric, or hydrochloric acid, or of certain salts, such as mercuric chloride and lead acetate. Metaphosphoric acid instantly precipitates albumen from its solutions. Orthophosphoric acid, acetic and lactic acids, form no precipitates with albumen, neither does common salt unless acetic acid be present.

570. FIBRIN.—When fresh blood is allowed to stand, it soon separates into a yellow liquid, called *serum*, and a red coagulum or clot. The clot contains the red corpuscles, the oxygen-carriers of the blood, imprisoned in a mass of insoluble albuminoid matter. By beating the fresh blood with a bunch of twigs or an egg-beater, the mass of blood is prevented from coagulating, and the albuminoid matter, which is called fibrin, becomes attached to the beater in red flakes. By washing in a stream of water, the red corpuscles are washed out, and the fibrin remains as light-gray, elastic filaments. It is insoluble in water, but dissolves in very dilute alkaline solutions. Fibrin is formed by the union of two substances contained in the blood, whenever that liquid is kept at rest. Its spontaneous coagulation causes the cessation of bleeding from slight cuts and other small wounds.

The stiffening of the muscles which takes place soon after death, is due to the coagulation of a peculiar albuminoid matter, called *myosin*, which exists in solution in the muscular tissues. It is soluble in water containing 10 per cent. of salt, but is precipitated by a larger quantity : it is extracted by virtue of this property.

571. HEMOGLOBIN is a crystallizable matter which can be extracted from the red corpuscles of blood. It contains a small proportion of iron. Hemoglobin has the property of absorbing oxygen and forming an unstable compound from which the oxygen

P   *w*   29

escapes by exposure in a vacuum. It is probably by this property of the hemoglobin which they contain, that the red blood corpuscles are enabled to carry oxygen to all parts of the system. Hemoglobin will also absorb carbon monoxide, and when it has absorbed that gas it is incapable of combining with oxygen : this explains the poisonous effects of carbon monoxide on the system. Hydrogen sulphide reduces hemoglobin,—that is, removes its oxygen ; and we can so understand the injurious action of any quantities of this gas.

572. CASEIN, MILK.—Milk is a dilute solution of lactose or milk sugar and a small quantity of mineral salts, in which are suspended very small fat globules, and, either suspended or in solution, an albuminoid matter called casein. The specific gravity of milk is about 1030. After standing for several hours, the greater number of the fat globules come to the surface, constituting the cream ; cream, however, contains some lactose, casein, and salts. Its composition varies greatly, as may be seen from the following results of the analysis of three samples :

	I.	II.	III.
Water	72.2	66.36	50
Fat	19.0	18.87	43.9
Casein, lactose, salts	8.8	14.77	6.1

The following is the composition of an average sample of cow's milk, but it must be borne in mind that no two samples will probably have the same composition. In one thousand parts, this milk contained 128 parts of solid matter, constituted as follows :

Casein	35.70
Butter	33.80
Lactose and salts	58.50

When an acid is added to milk, a thick deposit of coagulated casein is formed. This same substance is formed when milk naturally becomes sour by the formation of lactic and acetic acids. During the first stages of the souring of milk, lactic acid is produced by the fermentation of the lactose, but, unless a large quantity of lactic acid be present, this alone will not coagulate the casein ; as soon, however, as acetic acid begins to form, the casein

becomes insoluble and the milk thickens. Casein closely resembles insoluble albumen; it dissolves in dilute solutions of the
alkaline hydrates and carbonates, and it is probably in solution in
fresh milk, for that liquid has an alkaline reaction. Cheese is
modified casein.

573. GELATIN.—When bones are immersed in hydrochloric
acid, the mineral matter, consisting principally of calcium phosphate and carbonate, is dissolved, and a semi-transparent, elastic
mass is obtained, retaining the form of the bone. This body is
insoluble in cold water, but by long boiling it dissolves, and, on
cooling, the solution sets in a transparent jelly. This substance
is gelatin, or glue. It is not peculiar to the bones, but exists also
in certain other tissues, particularly in the skin, and in the swim-
ming-bladders of fishes. The best gelatin is obtained from the
swimming-bladder of the sturgeon; it is called fish-glue. Very
little is known regarding the difference between gelatin and the
substances from which it is derived.

Dry gelatin occurs in transparent or translucid sonorous sheets,
whose color varies from colorless to brown, according to the purity.
It swells in water, but does not dissolve until the liquid is heated.
Its solution is precipitated by alcohol, but not by acids, with the
exception of tannic acid, with which it forms an insoluble compound. The tanning of skins and hides depends on the formation
of this compound in the body of the skin, which is so converted
into leather.

574. By the processes of digestion, the vegetable and animal
matters which serve as food are converted into substances which
can be assimilated or made part of our bodies. These processes
begin in the mouth, where the starchy substances encounter in
the saliva a peculiar unorganized ferment called *ptyalin*, which is
capable of transforming them into soluble glucose. Ptyalin is
probably identical with diastase, which is formed during the germination of grain. In the stomach, the conversion of starch into
glucose continues, and the albuminoid matters are converted into
soluble bodies called peptones by another ferment, *pepsin*, contained,

together with a little hydrochloric acid, in the gastric juice. This ferment exists in rennet, obtained from the stomach of the calf, and used in the manufacture of cheese. The peptones appear to be formed by the hydration of the albuminoid bodies, and in the system they are probably converted into all the varieties of albuminoid tissue. As the food passes from the stomach it encounters in the small intestines other ferments, by which the fatty matters are emulsified and rendered capable of being absorbed and passing into the blood, by which they are carried and deposited where needed in the system.

The slow combustion by which life is sustained results in the oxidation of the tissues, and the removal of the matters no longer useful. This oxidation is not accomplished in one operation, but in several stages, during which many compounds intermediate between the albuminoid and fatty bodies, and the carbon dioxide, water, and nitrogen which would result from their complete combustion, are formed. We have seen that a great part of the carbon and hydrogen is indeed removed as carbon dioxide and water, but the salts are in great part eliminated unchanged by the urine and the perspiration. The nitrogen is excreted principally as urea, phosphorus as sodium acid phosphate, sulphur as sodium sulphate, etc. A small part of the nitrogen of the system is excreted in forms intermediate between the albuminoid bodies and urea. Among the more important of these is uric acid, $C^5H^4N^4O^3$, a compound forming a small proportion of human urine, and existing in large quantity in the solid urine of birds and reptiles.

In all these processes there is comparatively little that we can understand. We know only that they all result in a transfer of energy,—that in living matter the chemical energy is converted into the energy of life ; and we can comprehend only the beginning and the end of the phenomenon,—the forms of matter which are capable of organization, and the products of the disorganization without which life could not continue.

APPENDIX.

CRYSTALLOGRAPHY.

A crystal is a natural polyhedron ; that is, a solid bounded by plane surfaces or faces. The greater number of solid chemical substances form more or less perfect crystals whenever a certain freedom of motion is communicated to their molecules, so that these molecules may arrange themselves without interference. Such freedom of motion may be given to the molecules :

1. By dissolving the solid in any liquid by which it is not altered, and allowing a hot saturated solution to cool slowly, or by the spontaneous evaporation of the solvent if the solid be equally soluble at all temperatures. Potassium chlorate, lead iodide, potassium nitrate, and alum may be crystallized from water by the first method; common salt by the second.

2. By melting the solid, and decanting the still liquid portion after a crust has formed on its surface. Sulphur and bismuth may be so crystallized.

3. By subliming the solid, and allowing the vapor to condense very slowly. In this manner fine crystals of iodine and camphor may be obtained.

4. By a chemical reaction in which the crystallizable substance is formed in a medium in which it is insoluble. Platino-potassium chloride and potassium acid tartrate are so formed in microscopic crystals.

A solid which manifests no tendency to become crystalline is said to be *amorphous*.

For convenience' sake, crystals are classified in six systems, and each system is characterized by a set of axes, which are imaginary straight lines passing through the centre of the crystal and joining opposite solid angles or the centres of opposite faces or edges. The forms belonging to any one system may be derived from each other by cutting off the edges or angles by plane surfaces; in all such derivatives the imaginary axes must remain untouched.

1. THE ISOMETRIC SYSTEM has three equal axes at right angles to each other. The type of the system is the cube (*a*), in which the axes join the centres of opposite faces. The regular tetrahedron (*b*) is derived from the cube by cutting off every other solid angle by a plane which passes through the three adjacent angles. The regular octahedron (*c*) is formed by cutting off in the same manner all the solid angles. The rhombic dodecahedron (*d*) results when all the edges are cut off by planes which meet in the centres of the faces.

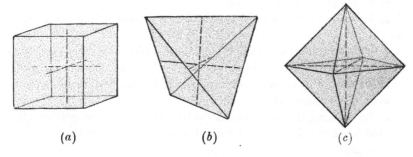

(*a*)   (*b*)   (*c*)

2. THE QUADRATIC SYSTEM has three axes at right angles to each other, and two of them are equal The type is the right square prism (*e*), in which the axes join the centres of opposite faces. This system includes the right square octahedron (*f*).

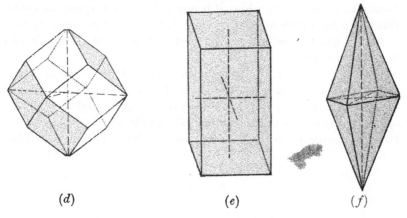

(*d*)   (*e*)   (*f*)

3. THE ORTHORHOMBIC SYSTEM is characterized by three unequal axes at right angles to each other. The type is the right rectangular prism (*g*), in which the axes join the centres of opposite faces. By cutting off the vertical edges until the new plane faces meet, the right

rhombic prism is obtained. Here the horizontal axes join the centres of opposite edges, while the vertical axis joins opposite faces. The right rhombic octahedron (*h*) belongs to this system.

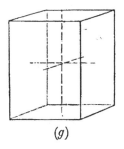

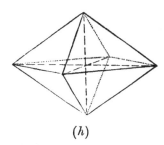

(*g*)   (*h*)

4. THE HEXAGONAL SYSTEM has four axes; three are in the same plane, and would join the opposite angles of a regular hexagon. The fourth is at right angles to these three, and passes through their point of intersection. The hexagonal prism (*i*), hexagonal pyramid, and rhombohedron (*j*) belong to this system.

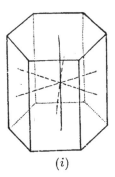

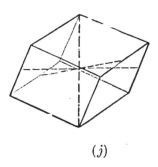

(*i*)   (*j*)

5. THE CLINORHOMBIC SYSTEM has three axes, all of which are unequal; two of them are inclined to each other and in the same plane, but both are at right angles to the third, which passes through their point of intersection. The right rhomboidal prism, the oblique rectangular prism, oblique rhombic prism, and oblique rhombic octahedron are of this system.

6. THE ANORTHIC SYSTEM has three unequal axes, neither of which is at right angles with another. The oblique rhomboidal prism is the type of this system.

All of the crystalline forms named are primitive forms; they may be modified in many manners, but the characteristic axes are not affected by the modifications.

A substance which crystallizes in forms belonging to two different systems is said to be *dimorphous*. Such a body is sulphur.

Two different substances whose crystals are of precisely the same form are said to be *isomorphous*. The chlorides, bromides, and iodides of potassium and sodium are isomorphous: the alums are an excellent example of isomorphous compounds.

HINTS FOR THE PREPARATION OF EXPERIMENTS.

Unless the floor of a class-room inclines to the front, the table on which apparatus is arranged and experiments performed should be quite high. A table closed at the front and ends, and, with the exception of a few drawers, entirely open at the back, is the most convenient, as a large quantity of apparatus may be kept ready for use under it. If possible, the table should be furnished with gas- and water-pipes having several taps, and these are most convenient if run along under the back edge, the nozzles of the taps being flush with the back. A water-supply naturally necessitates a sink for the waste-water, and this may well be set in the table, being provided with a cover pierced with holes for the drainage of funnels, condensing apparatus, etc., when the whole surface of the table is required.

A pneumatic trough sunk in the table is very convenient for the teacher, but is very unsatisfactory to a class of beginners, who desire to see every process of manipulation. It is therefore much better to use a trough entirely of glass, and nothing answers the purpose better than a strong aquarium trough: this may be fitted with a movable shelf, suspended from the edges of the frame, and is satisfactory to teacher and pupils.

All the apparatus and chemicals required for ordinary experiments should be kept as close at hand as possible, and ready for use; it is advantageous, when possible, to use the same set of apparatus only for the same experiment. In this manner the time required for mechanical preparation may be reduced to a minimum, and in case of accident any part of an apparatus may be quickly replaced.

It is true economy to provide a large stock of flasks, from half-litre to two litres capacity, glass tubing of all sizes, funnel-tubes, beakers, and rubber tubing and corks. Rubber corks may be obtained pierced with one, two, or three holes, and though the expense of a stock of these corks—say half a dozen of each useful size—is considerable at first, the economy of time which would otherwise be required for fitting and boring corks, amply repays the outlay. The rubber corks made by Galante, 2 Rue de l'Ecole de Médecine, Paris, are the best, and with proper care may be used several years.

If rubber corks cannot be obtained, ordinary cork may be bored by cork-borers or by a red-hot pointed iron wire. The hole may then be enlarged to any size by rat-tail files. Corks should be adapted by cutting and filing to fit hermetically without the use of sealing-wax or luting of any kind : a sharp shoemaker's knife, a good rasp, and several flat files are required for this purpose.

Glass flasks and porcelain apparatus should not be heated by a naked flame, but should be placed on wire gauze, which may be cut into squares of the required size, and its use enables the immediate application of heat to glass-ware which might be cracked if heated without precaution.

Naturally, gas is the best and most convenient heating agent, but when this cannot be obtained it is in most cases better to use an oil stove than a spirit-lamp. There are many patterns of oil stoves in which the heated gases from several flames may be entirely employed in very little space, and with care they will not deposit smoke on apparatus heated over them.

A blast-lamp is indispensable for the operations of glass-working, and, unless gas be at hand, a kerosene lamp with a large loose wick may be employed. The nozzle of the blast may be made of glass tube, drawn out to the required opening, and this may be supported in a cork which can be raised or lowered on a vertical iron rod. When gas is employed for the blast-lamp, a single flame is much more satisfactory in the laboratory than the double flame generally employed by glass-blowers in this country. The most convenient and cheapest bellows is that in which the reservoir is a caoutchouc bag. It is durable, and can be adapted to all laboratory purposes in which a blast is required.

Glass tubes are cut by notching them *once* with the smallest-sized triangular file, firmly grasping each end between the thumb and finger, close to the notch, and bending from the filed side. The sharp ends may then be rounded by holding them in a flame until the glass begins to soften.

The art of glass-blowing can be acquired only by practice ; a pound of glass tubes and a few hours at the lamp are of more value than any instruction. The beginner must learn to regulate the size of his flame according to the work to be done, to heat the glass evenly, to observe the moment when it is at the proper temperature, and to remove it from the flame before bending, drawing, or blowing.

The glass must be hotter for bending than for drawing, and hotter for blowing than for bending.   In making a bend, the whole length of the bend should be uniformly heated by holding it in a slanting position through the flame, and rotating it slowly.   The bend should not be a sharp angle, but a regular curve.   To blow a bulb, the glass should be somewhat thickened by softening the portion and gently pressing the ends towards each other.   If the bulb is to be on the end of a tube, the latter will be sealed and shortened by the cohesion of the softened glass, and the bulb may then be blown in the thick end.   Practice and patience are the best teachers.

It must be remembered that the pupil's interest in the science will depend on his comprehension of the facts, and all apparatus intended for class-demonstration should be large, so that each part can be seen and understood by every pupil.   In addition, the apparatus should be neatly arranged : clumsy contrivances may accomplish results, but they often fail in the more important object of exhibiting the process to which the result is due.

INDEX.

THE END.

THE READERS FOR YOUR SCHOOLS.

LIPPINCOTT'S
Popular Series of Readers.
By MARCIUS WILLSON.

THIS SERIES OF READERS EMBRACES SIX BOOKS, AS FOLLOWS:

FIRST READER. With Illustrations.* 96 pages. 12mo. Half bound. 20 cents.†

SECOND READER. With Illustrations. 160 pages. 12mo. Half bound. 33 cents.†

THIRD READER. With Illustrations. 228 pages. 12mo. Half bound. 44 cents.†

FOURTH READER. With Illustrations. 334 pages. 12mo. Half bound. 60 cents.†

FIFTH READER. With Illustrations. 480 pages. 12mo. Cloth sides. 90 cents.†

SIXTH READER. With Frontispiece. 544 pages. 12mo. Cloth sides. $1.00.†

They combine the greatest possible interest with appropriate instruction.

They contain a greater variety of reading matter than is usually found in School Readers.

They are adapted to modern methods of teaching.

They are natural in method, and the exercises progressive.

They stimulate the pupils to think and inquire, and therefore interest and instruct.

They teach the principles of natural and effective reading.

The introduction of SCRIPT EXERCISES is a new feature, and highly commended by teachers.

The LANGUAGE LESSONS accompanying the exercises in reading mark a new epoch in the history of a Reader.

The ILLUSTRATIONS are by some of the best artists, and represent both home and foreign scenes.

"No other series is so discreetly graded, so beautifully printed, or so philosophically arranged." —*Albany Journal.*

"We see in this series the beginning of a better and brighter day for the reading classes."—*New York School Journal.*

"The work may be justly esteemed as the beginning of a new era in school literature."—*Baltimore News.*

"In point of interest and attractiveness the selections certainly surpass any of the kind that have come to our knowledge." — *The Boston Sunday Globe.*

The unanimity with which the Educational Press has commended the Popular Series of Readers is, we believe, without a parallel in the history of similar publications, and one of the best evidences that the books meet the wants of the progressive teacher.

WS - #0031 - 201023 - C0 - 229/152/20 - PB - 9781330259399 - Gloss Lamination